Fundamentals of Quantum Chemistry

Second Edition

The Complimentary Science Series in an introductory, interdisciplinary, and relatively inexpensive series of paperbacks for science enthusiasts. These titles cover topics that are particularly appropriate for self-study although they are often used as complementary texts to supplement standard discussion in textbooks. They are deliberately unburdened by excessive pedagogy, which is distracting to many readers, and avoid the often plodding treatment in introductory texts. The series was conceived to fill the gaps in the literature between conventional textbooks and monographs by providing real science at an accessible level, with minimal prerequisites so that students at all stages can have expert insight into important and foundational aspects of current scientific thinking.

Many of these titles have strong interdisciplinary appeal, such as a chemist writing about applications of biology to physics, or vice versa, and all have a place on the bookshelves of literature laypersons. Potential authors are invited to contact our editorial office at www.academicpressbooks.com.

Complimentary Science Series

ELSEVIER
ACADEMIC
PRESS

Earth Magnetism
Wallace Hall Campbell

Physics in Biology and Medicine, 2nd Edition
Paul Davidovits

Mathematics for Physical Chemistry, 2nd Edition
Robert Mortimer

The Physical Basis of Chemistry, 2nd Edition
Warren S. Warren

Introduction to Relativity
John B. Kogut

Chemistry Connections: The Chemical Basis of Everyday Phenomena, 2nd Edition
Kerry K. Karukstis and Gerald R. Van Hecke

www.academicpressbooks.com

Senior Editor, Sciences	Jeremy Hayhurst
Editorial Coordinator	Nora Donaghy
Senior Project Manager	Angela Dooley
Marketing Manager	Linda Beattie
Cover Design	G.B.D. Smith
Copyeditor	Charles Lauder, Jr.
Composition	Integra
Printer	Maple-Vail

This book is printed on acid-free paper. ∞

Academic Press
An imprint of Elsevier Science
525 B Street, Suite 1900, San Diego, California 92101-4495, USA
http://www.academicpress.com

Academic Press
84 Theobald's Road, London WC1X 8RR, UK
http://www.academicpress.com

Academic Press
200 Wheeler Road, Burlington, Massachusetts 01803, USA
www.academicpressbooks.com

Library of Congress Cataloging-in-Publication Data
A catalog record for this book is available from the Library of Congress

International Standard Book Number: 0-12-356771-8

PRINTED IN THE UNITED STATES OF AMERICA
03 04 05 06 9 8 7 6 5 4 3 2 1

Fundamentals of Quantum Chemistry

Second Edition

James E. House

Illinois State University

ELSEVIER
ACADEMIC
PRESS

Amsterdam Boston Heidelberg London New York Oxford
Paris San Diego San Francisco Singapore Sydney Tokyo

Contents

Preface to the Second Edition

This second edition of *Fundamentals of Quantum Chemistry* is an expansion of the successful first edition, which was published as *Fundamentals of Quantum Mechanics* (1998). My goal, then and now, was to provide a clear, readable presentation of the basic principles and application of quantum mechanical models for chemists while maintaining a level of mathematical completeness that enables the reader to follow the developments. The title has been changed to more accurately represent the book to a readership with a chemical rather than a physical specialization. Of course, much of the material is equally applicable to both audiences, and the complete contents of the first edition are retained herein.

The second edition differs from the first in several ways.

1. A new chapter on molecular orbital calculations (extended Hückel and self-consistent field), which introduces some of the basic ideas and terminology of the topic, has been included.

2. Several new topics, as sections or part of sections, have been included. These include the photoelectric effect, the perturbation treatment of the helium atom, orbital symmetry and chemical reactions, and molecular term symbols.

3. A significant number of additional figures and minor improvements to existing figures have been added.

4. A significant number of new exercises have been included.

5. Answers are now provided for selected problems at the back of the book.

6. Last but not least, the entire text has been carefully and extensively edited to increase the clarity of the presentation and to correct minor errors.

I believe that these changes will enhance the relevance of the book for a wide range of readers.

It is a pleasure to acknowledge the outstanding guidance and support of Jeremy Hayhurst, Angela Dooley, and Nora Donaghy of Academic Press. Working with them again has been a pleasant experience that I hope to repeat. As in my other book writing ventures, the support and encouragement from my wife, Kathleen A. House, have been invaluable.

Preface to the First Edition

A knowledge of quantum mechanics is indispensable to understanding many areas of the physical sciences. In addition to courses dealing specifically with quantum mechanics, some coverage is devoted to quantum mechanics in many other courses to provide background for the study of certain specific topics. An enormous number of advanced texts in quantum mechanics and quantum chemistry exist for the advanced student or specialist. However, there are few books that deal with quantum mechanics on an elementary level to provide the type of survey needed by nonspecialists to understand the basis of experiments and theories in their fields. My experience in teaching several of these peripheral courses leads me to believe that many students at several levels need some exposure to the main ideas of quantum mechanics. I am also convinced that it is profitable for many students to obtain that exposure from a book that is not intended for study of the subject at an advanced level. Because of this, I have tried to write the book that I wish I had had at the beginning of my study of quantum mechanics.

In my teaching, I encounter a large number of students in chemistry at the undergraduate and M.S. levels who need to review basic quantum mechanics. By actual survey, the vast majority of these students stopped their preparation in mathematics after the required two semesters of calculus. This is typical of students who do not plan to take more specialized courses in quantum mechanics and quantum chemistry. The situation is somewhat similar for students at certain levels in biological sciences, physics, and engineering. The purpose of this book is to provide a minimal background in quantum mechanics quickly and concisely for anyone who needs such a survey. It should also be suitable as a review of the subject for those who are no longer students but who need (or want!) to know some quantum mechanics.

With this audience in mind, this book has been kept to a level that makes it usable by persons of limited background in mathematics. It is presumed that the reader is familiar with basic physics and calculus, but no other background is assumed. In fact, this is one of the intended strengths of this volume, and a few mathematical topics are included in considerable detail to bring the reader along with elementary topics in differential equations, determinants, etc. In this sense, the book is a tool for self-teaching. Of course, no small book can cover quantum mechanics in either depth or breadth. The choice of topics was based on the applicability and relevance of the material to the larger fields of the physical sciences. Much of applied quantum mechanics is based on the treatment of several model systems (particle in a box, harmonic oscillator, rigid rotor, barrier penetration, etc.). These models form the content of much of the survey of quantum mechanics presented here.

After working through this book, the reader will have some familiarity with most of the important models of quantum mechanics. For those whose needs exceed the presentation here or whose appetite for quantum mechanics has been whetted, references are included at the end of each chapter.

It is hoped that this book will meet the needs of a wide audience. First, it should be a useful supplement for a variety of courses in the physical sciences. Second, it should serve as a tool for self-study and review by persons who have ended their formal education. Finally, this book should be a useful lead-in for students (especially those of limited mathematical background) preparing to study the more advanced works in the field. As stated earlier, my aim was to write the book that I wish I had had to start the learning of quantum mechanics.

Debra Feger-Majewski, Dustin Mergott, Sara McGrath, Anton Jerkovich, Ovette Villavicencio, Matt Lewellen, and Jeff Zigmant used some of this material in a preliminary form and made many useful suggestions. The reviews of the manuscript provided by Dr. Clarke W. Earley and Dr. Earl F. Pearson have contributed greatly to this book. Further, David Phanco, Garrett Brown, Jacqueline Garrett, and Michael Remener have made the development and production of this book a pleasant and rewarding experience. Finally, the patience and understanding of my wife, Dr. Kathleen A. House, during the writing of this book are gratefully acknowledged. Her assistance with graphics production and her careful reading of the entire manuscript have contributed greatly to this book, and her encouragement since the inception of this project has helped make yet another dream come true.

Chapter 1

The Early Days

Quantum mechanics is a branch of science that deals with atomic and molecular properties and behavior of matter on a microscopic scale. While thermodynamics may be concerned with the heat capacity of a gaseous sample, quantum mechanics is concerned with the specific changes in rotational energy states of the molecules. Chemical kinetics may deal with the rate of change of one substance into another, but quantum mechanics is concerned with the changes in vibrational states and structures of the reactant molecules as they are transformed. Quantum mechanics is also concerned with the spins of atomic nuclei and the populations of excited states of atoms. Spectroscopy is based on changes of quantized energy levels of several types. Quantum mechanics is thus seen to merge with many other areas of modern science.

A knowledge of the main ideas and methods of quantum mechanics is important for developing an understanding of branches of science from nuclear physics to organic chemistry. This book attempts to develop that familiarity for persons from all of the sciences.

The modern applications of quantum mechanics have their roots in the developments of physics around the turn of the century. Some of the experiments, now almost 100 years old, provide the physical basis for interpretations of quantum mechanics. The names associated with much of this early work (e.g., Planck, Einstein, Bohr, de Broglie) are legendary in the realm of physics. Their elegant experiments and theories now seem almost commonplace to even beginning students, but these experiments were at the forefront of scientific development in their time. It is appropriate, therefore, for this book to begin with a brief review of a few of the most important of these early studies.

1.1 Blackbody Radiation

When an object is heated to incandescence it emits electromagnetic radiation. The nature of the object determines to some extent the type of radiation that is emitted, but in all cases a range or distribution of radiation is produced. It is known that the best absorber of radiation is also the best emitter of radiation. The best absorber is a so-called "blackbody," which absorbs all radiation and from which none is reflected. If we heat this blackbody to incandescence, it will emit a whole range of electromagnetic radiations whose energy distributions depend on the temperature to which the blackbody is heated. Early attempts to explain the distribution of radiation using the laws of classical physics were not successful. In these attempts it was assumed that the radiation was emitted because of vibrations or oscillations within the blackbody. These attempts failed to explain the position of the maximum that occurs in the distribution of radiation; in fact, they failed to predict the maximum at all.

Since radiation having a range of frequencies (ν, Greek "nu") is emitted from the blackbody, theoreticians tried to obtain an expression that would predict the relative intensity (amount of radiation) of each frequency. One of the early attempts to explain blackbody radiation was made by W. Wien. The general form of the equation that Wien obtained is

$$f(\nu) = \nu^3 g(\nu/T), \tag{1.1}$$

where $f(\nu)$ is the amount of energy of frequency ν emitted per unit volume of the blackbody and $g(\nu/T)$ is some function of ν/T. This result gave fair agreement with the observed energy distribution at longer wavelengths but did not give agreement at all in the region of short wavelengths. Another relationship obtained by the use of classical mechanics is the expression derived by Lord Rayleigh,

$$f(\nu) = \frac{8\pi\nu^3}{c^3}kT, \tag{1.2}$$

where c is the velocity of light (3.00×10^8 m/s) and k is Boltzmann's constant, 1.38×10^{-16} erg/molecule.

Another expression was found by Rayleigh and Jeans and predicts the shape of the energy distribution as a function of frequency, but only in the region of short wavelengths. The expression is

$$f(\nu) = \frac{8\pi \nu^3}{c^3}\left(\frac{hkT}{h\nu}\right) = \frac{8\pi \nu^2 kT}{c^3}. \tag{1.3}$$

Therefore, the Wien relationship predicted the intensity of high-ν radiation, and the Rayleigh–Jeans law predicted the intensity of low-ν radiation emitted from a blackbody. Neither of these relationships predicted a distribution of radiation that goes through a maximum at some frequency with smaller amounts emitted on either end of the spectrum.

The problem of blackbody radiation was finally explained in a satisfactory way by Max Planck in 1900. Planck still assumed that the absorption and emission of radiation arose from some sort of oscillators. Planck made a fundamental assumption that only certain frequencies were possible for the oscillators instead of the whole range of frequencies that were predicted by classical mechanics. The permissible frequencies were presumed to be some multiple of a fundamental frequency of the oscillators, ν_0. The allowed frequencies were then $\nu_0, 2\nu_0, 3\nu_0, \ldots$. Planck also assumed that energy needed to be absorbed to cause the oscillator to go from one allowed frequency to the next higher one and that energy was emitted as the frequency dropped by ν_0. He also assumed that the change in energy was proportional to the fundamental frequency, ν_0. Introducing the constant of proportionality, h,

$$E = h\nu_0, \tag{1.4}$$

where h is Planck's constant, 6.63×10^{-27} erg s or 6.63×10^{-34} J s. The average energy per oscillator was found to be

$$\langle E \rangle = \frac{h\nu_0}{\left(e^{h\nu_0/kT} - 1\right)}. \tag{1.5}$$

Planck showed that the emitted radiation has a distribution given by

$$f(\nu) = \frac{8\pi \nu_0^3}{c^3}\langle E \rangle = \frac{8\pi \nu_0^3}{c^3}\frac{h\nu_0}{e^{h\nu_0/kT} - 1}. \tag{1.6}$$

This equation predicted the observed relationship between the frequencies of radiation emitted and the intensity.

The successful interpretation of blackbody radiation by Planck provided the basis for considering energy as being quantized, which is so fundamental to our understanding of atomic and molecular structure and our experimental methods for studying matter. Also, we now have the relationship between the frequency of radiation and its energy,

$$E = h\nu. \tag{1.7}$$

These ideas will be seen many times as one studies quantum mechanics and its application to physical problems.

1.2 The Line Spectrum of Atomic Hydrogen

When gaseous hydrogen is enclosed in a glass tube in such a way that a high potential difference can be placed across the tube, the gas emits a brilliant reddish-purple light. If this light is viewed through a spectroscope, the four major lines in the visible *line spectrum* of hydrogen are seen. There are other lines that occur in other regions of the electromagnetic spectrum that are not visible to the eye. In this visible part of the hydrogen emission spectrum, the lines have the wavelengths

$$\begin{aligned}
H_\alpha &= 6562.8\ \text{Å} = 656.28\ \text{nm (red)}\\
H_\beta &= 4961.3\ \text{Å} = 486.13\ \text{nm (green)}\\
H_\gamma &= 4340.5\ \text{Å} = 434.05\ \text{nm (blue)}\\
H_\delta &= 4101.7\ \text{Å} = 410.17\ \text{nm (violet).}
\end{aligned}$$

As shown in Figure 1.1, electromagnetic radiation is alternating electric and magnetic fields that are perpendicular and in phase. Planck showed that the energy of the electromagnetic radiation is proportional to the frequency, ν, so that

$$E = h\nu. \tag{1.8}$$

Since electromagnetic radiation is a transverse wave, there is a relationship between the wavelength, λ (Greek "lambda"), and the frequency, ν. Frequency is expressed in terms of cycles per unit time, but a "cycle" is simply

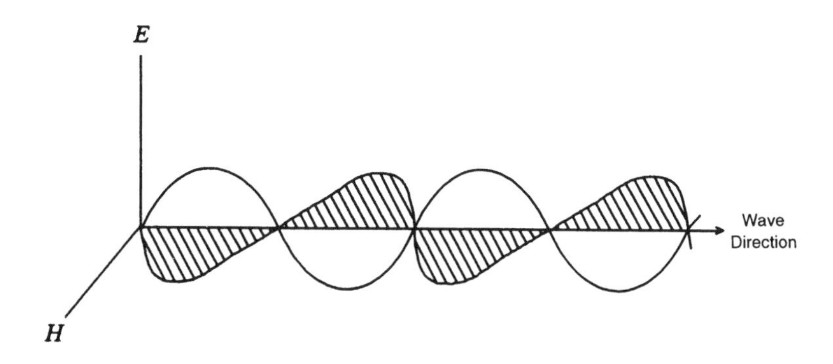

Figure 1.1 ▶ Electromagnetic radiation.

a count, which carries no units. Therefore, the dimensions of frequency are "cycles"/time or 1/time. The wavelength is a distance so its dimension is distance (or length). The product of wavelength and frequency is

$$\lambda\nu = \text{distance} \times \frac{1}{\text{time}} = \frac{\text{distance}}{\text{time}} = \text{velocity} = v. \qquad (1.9)$$

In the case of electromagnetic radiation, the velocity of light is c, which is 3.00×10^{10} cm/s. Therefore, $\lambda\nu = c$ and

$$E = h\nu = \frac{hc}{\lambda}. \qquad (1.10)$$

In 1885, Balmer discovered an empirical formula that would predict the preceding wavelengths. Neither Balmer nor anyone else knew why this formula worked, but it did predict the wavelengths of the lines accurately. Balmer's formula is

$$\lambda(\text{cm}) = 3645.6 \times 10^{-8}\left(\frac{n^2}{n^2 - 2^2}\right). \qquad (1.11)$$

The constant 3645.6×10^{-8} has units of centimeters and n represents a whole number larger than 2. Using this formula, Balmer was able to predict the existence of a fifth line. This line was discovered to exist at the boundary between the visible and the ultraviolet regions of the spectrum. The measured wavelength of this line agreed almost perfectly with Balmer's prediction.

Balmer's empirical formula also predicted the existence of other lines in the infrared (IR) and ultraviolet (UV) regions of the spectrum of hydrogen. These are as follows:

Lyman series: $n^2/(n^2 - 1^2)$, where $n = 2, 3, \ldots$(1906–1914, UV)

Paschen series: $n^2/(n^2 - 3^2)$, where $n = 4, 5, \ldots$(1908, IR)

Brackett series: $n^2/(n^2 - 4^2)$, where $n = 5, 6, \ldots$(1922, IR)

Pfund series: $n^2/(n^2 - 5^2)$, where $n = 6, 7, \ldots$(1924, IR).

Balmer's formula can be written in terms of 1/wavelength and is usually seen in this form. The equation becomes

$$\frac{1}{\lambda} = R\left(\frac{1}{2^2} - \frac{1}{n^2}\right), \qquad (1.12)$$

where R is a constant known as the *Rydberg constant*, 109,677.76 cm^{-1}. The quantity $\frac{1}{\lambda}$ is called the *wave number* and is expressed in units of centimeters^{-1}(cm^{-1}). The empirical formulas can be combined into a general form

$$\bar{\nu} = \frac{1}{\lambda} = R\left(\frac{1}{2_1^2} - \frac{1}{n_2^2}\right). \tag{1.13}$$

When $n_1 = 1$ and $n_2 = 2, 3, 4, \ldots,$ the Lyman series is predicted. For $n_1 = 2$ and $n_2 = 2, 3, 4, \ldots,$ the Balmer series is predicted, and so on. Other empirical formulas that correlated lines in the spectra of other atoms were found, but the same constant R occurred in these formulas. At the time, no one was able to relate these formulas to classical electromagnetic theory.

1.3 The Bohr Model for the Hydrogen Atom

It is not surprising that the spectrum of the hydrogen atom was the first to be explained since it is the simplest atom. E. Rutherford had shown in 1911 that the model of the atom is one in which a small but massive positive region is located in the center of the atom and the negative region surrounds it. Applying this model to the hydrogen atom, the single proton is located as the nucleus while a single electron surrounds or moves around it. N. Bohr incorporated these ideas into the first dynamic model of the hydrogen atom in 1913, supposing the electron to be governed by the laws of classical or Newtonian physics. There were problems, however, that could not be answered by the laws of classical mechanics. For example, it had been shown that an accelerated electric charge radiates electromagnetic energy (as does an antenna for the emission of radio frequency waves). To account for the fact that an atom is a stable entity, it was observed that the electron must move around the nucleus in such a way that the centrifugal force exactly balances the electrostatic force of attraction between the proton and the electron. Since the electron is moving in some kind of circular orbit, it must constantly undergo acceleration and should radiate electromagnetic energy by the laws of classical physics.

Because the Balmer series of lines in the spectrum of atomic hydrogen had been observed earlier, physicists attempted to use the laws of classical physics to explain a structure of the hydrogen atom that would give rise to these lines. It was recognized from Rutherford's work that the nucleus of an atom was surrounded by the electrons and that the electrons must be moving. In fact, no system of electric charges can be in equilibrium at rest.

While the electron in the hydrogen atom must be moving, there is a major problem. If the electron circles the nucleus, it is undergoing a constant

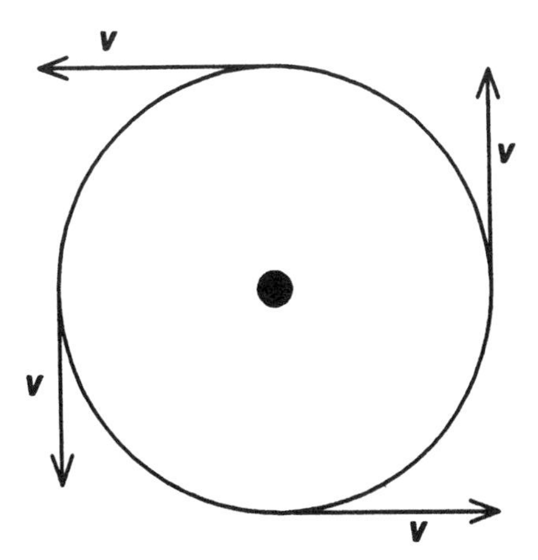

Figure 1.2 ▶ The change in velocity vector for circular motion.

change in direction as shown in Figure 1.2. Velocity has both a magnitude and a direction. Changing direction constitutes a change in velocity, and the change in velocity with time is acceleration. The laws of classical electromagnetic theory predict that an accelerated electric charge should radiate electromagnetic energy. If the electron did emit electromagnetic energy, it would lose part of its energy, and as it did so it would spiral into the nucleus and the atom would collapse. Also, electromagnetic energy of a *continuous* nature would be emitted, not just a few lines.

Bohr had to assume that there were certain orbits (the "allowed orbits") in which the electron could move without radiating electromagnetic energy. These orbits were characterized by the relationship

$$mvr = n\frac{h}{2\pi}, \tag{1.14}$$

where m is the mass of the electron, v is its velocity, r is the radius of the orbit, h is Planck's constant, and n is an integer, 1, 2, 3, ... Since n is a whole number, it is called a *quantum number*. This enabled the problem to be solved, but no one knew why this worked. Bohr also assumed that the emitted spectral lines resulted from the electron falling from an orbital of higher n to one of lower n. Figure 1.3 shows the forces acting on the orbiting electron.

The magnitudes of these forces must be equal for an electron to be in a stable orbit, so if e is the electron charge,

$$\frac{mv^2}{r} = \frac{e^2}{r^2}. \tag{1.15}$$

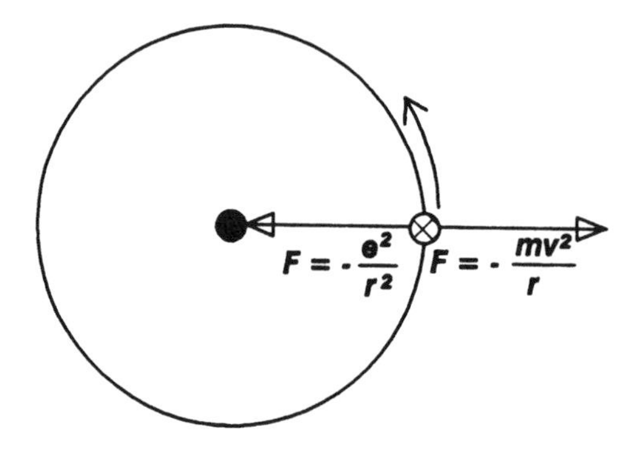

Figure 1.3 ▶ The forces on the electron in circular motion in a hydrogen atom.

Therefore, solving this equation for v gives

$$v = \sqrt{\frac{e^2}{mr}}. \tag{1.16}$$

From the Bohr assumption that

$$mvr = n\frac{h}{2\pi}, \tag{1.17}$$

solving for v gives

$$v = \frac{nh}{2\pi mr}. \tag{1.18}$$

Therefore, equating the two expressions for velocity gives

$$\sqrt{\frac{e^2}{mr}} = \frac{nh}{2\pi mr}. \tag{1.19}$$

Solving for r we obtain

$$r = \frac{n^2h^2}{4\pi^2me^2}. \tag{1.20}$$

This relationship shows that the radii of the allowed orbits increase as n^2 (h, m, and e are constants). Therefore, the orbit with $n = 2$ is four times as large as that with $n = 1$; the orbit with $n = 3$ is nine times as large as that with $n = 1$, etc. Figure 1.4 shows the first few allowed orbits drawn to scale. The units on r can be found from the units on the constants since e is measured in *electrostatic units* (esu) and an esu is in $\mathrm{g^{1/2}\ cm^{3/2}\ s^{-1}}$. Therefore,

$$\frac{[(\mathrm{g\ cm^2/s^2})\mathrm{s}]^2}{\mathrm{g}(\mathrm{g^{1/2}cm^{3/2}/s})^2} = \mathrm{cm}.$$

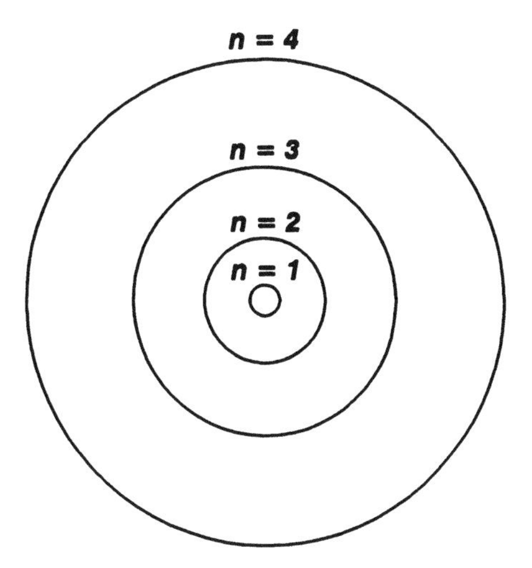

Figure 1.4 ▶ The first four orbits in the hydrogen atom drawn to scale.

The total energy is the sum of the electrostatic energy (potential) and the kinetic energy of the moving electron (total energy = kinetic + potential):

$$E = \frac{1}{2}mv^2 - \frac{e^2}{r}. \tag{1.21}$$

From equating the magnitudes of the centripetal and centrifugal forces, we know that

$$\frac{mv^2}{r} = \frac{e^2}{r^2} \tag{1.22}$$

or

$$mv^2 = \frac{e^2}{r}. \tag{1.23}$$

Multiplying both sides of this equation by 1/2 gives

$$\frac{1}{2}mv^2 = e^2/2r, \tag{1.24}$$

The left-hand side of Eq. (1.24) is simply the kinetic energy of the electron, and substituting this into Eq. (1.21) yields

$$E = \frac{e^2}{2r} - \frac{e^2}{r} = -\frac{e^2}{2r}. \tag{1.25}$$

We found earlier in Eq. (1.20) that

$$r = \frac{n^2h^2}{4\pi^2me^2},$$

and when we substitute this result for r in Eq. (1.25) we obtain

$$E = -\frac{e^2}{2r} = -\frac{e^2}{2\left(n^2h^2/4\pi^2me^2\right)} = -\frac{2\pi^2me^4}{n^2h^2}. \quad (1.26)$$

From this equation, we see that the energy of the electron in the allowed orbits varies *inversely* as n^2. Note also that the energy is *negative* and becomes less negative as n increases. At $n = \infty$ (complete separation of the proton and electron), $E = 0$ and there is no binding energy of the electron to the nucleus. The units for E in the previous equation depend on the units used for the constants. If h is in ergs seconds, the mass of the electron is in grams, and the charge on the electron, e, is in esu, we have

$$E = \frac{\mathrm{g(g^{1/2}cm^{3/2}/s)^4}}{\left[\mathrm{(g\ cm^2/s^2)\ s}\right]^2} = \frac{\mathrm{g\ cm^2}}{\mathrm{s^2}} = \mathrm{erg}. \quad (1.27)$$

We can then use conversion factors to obtain the energy in any other desired units.

If we write the expression for energy in the form

$$E = -\frac{1}{n^2}\frac{2\pi^2me^4}{h^2}, \quad (1.28)$$

we can evaluate the collection of constants when $n = 1$ to give -2.17×10^{-11} erg and assign various values for n to evaluate the energies of the allowed orbits:

$$n = 1,\ E = -21.7 \times 10^{-12}\ \mathrm{erg}$$
$$n = 2,\ E = -5.43 \times 10^{-12}\ \mathrm{erg}$$
$$n = 3,\ E = -2.41 \times 10^{-12}\ \mathrm{erg}$$
$$n = 4,\ E = -1.36 \times 10^{-12}\ \mathrm{erg}$$
$$n = 5,\ E = -0.87 \times 10^{-12}\ \mathrm{erg}$$
$$n = 6,\ E = -0.63 \times 10^{-12}\ \mathrm{erg}$$
$$n = \infty,\ E = 0.$$

Figure 1.5 shows an energy level diagram in which the energies are shown graphically to scale for these values of n. Note that the energy levels get closer together (converge) as the n value increases.

Causing the electron to be moved to a higher energy level requires energy because the positive and negative charges are held together by a

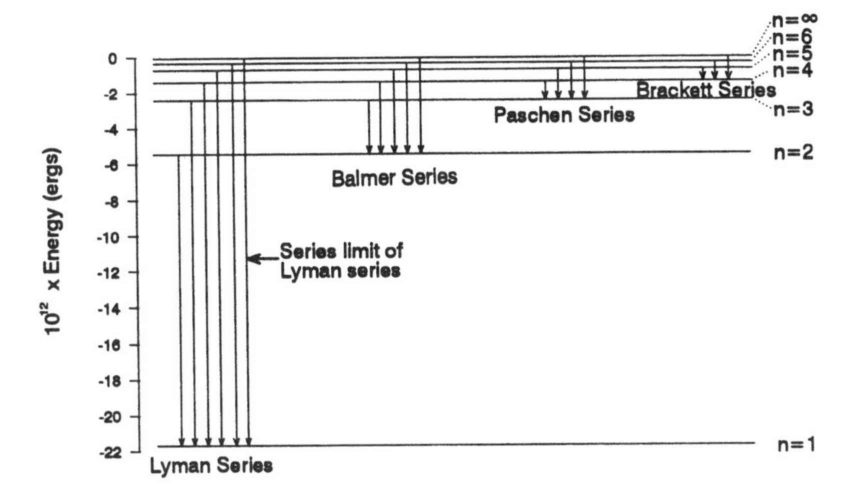

Figure 1.5 ▶ Energy level diagram for the hydrogen atom.

strong electrostatic force. Complete removal of the electron requires an amount of energy known as the *ionization potential* (or ionization energy) and corresponds to moving the electron to the orbital where $n = \infty$. The electron in the lowest energy state is held with an energy of -21.7×10^{-12} erg, and at $n = \infty$ the energy is 0. Therefore, the ionization potential for the H atom is 21.7×10^{-12} erg.

If we consider the energy difference between the $n = 2$ and the $n = 3$ orbits, we see that the difference is 3.02×10^{-12} erg. Calculating the wavelength of light having this energy, we find

$$E = h\nu = \frac{hc}{\lambda} \tag{1.29}$$

or

$$\begin{aligned}\lambda &= \frac{hc}{E} \\ &= \frac{(6.63 \times 10^{-27} \text{ erg s}) \times (3.00 \times 10^{10} \text{ cm/s})}{3.02 \times 10^{-12} \text{ erg}} \\ &= 6.59 \times 10^{-5} \text{ cm},\end{aligned} \tag{1.30}$$

which matches the wavelength of one of the lines in the Balmer series! Using the energy difference between the $n = 2$ and the $n = 4$ orbits leads to a wavelength of 4.89×10^{-5} cm, which matches the wavelength of another line in Balmer's series. Finally, the energy difference between the

$n = 2$ and the $n = \infty$ orbits leads to a wavelength of 3.66×10^{-5} cm, and this is the wavelength of the *series limit* of the Balmer series. It should be readily apparent that Balmer's series corresponds to light emitted as the electron falls from states with $n = 3, 4, 5, \ldots, \infty$, to the orbit with $n = 2$.

We could calculate the energies of lines emitted as the electron falls from orbits with $n = 2, 3, \ldots, \infty$, to the orbital with $n = 1$ and find that these energies match the lines in another observed spectral series, the *Lyman* series. In that case, the wavelengths of the spectral lines are so short that the lines are no longer in the visible region of the spectrum, but rather they are in the ultraviolet region. Other series of lines correspond to the transitions from higher h values to $n = 3$ (Paschen series, infrared), $n = 4$ (Brackett series, infrared), and $n = 5$ (Pfund series, far infrared) as the lower values.

The fact that the series limit for the Lyman series represents the quantity of energy that would be required to remove the electron suggests that this is one way to obtain the ionization potential for hydrogen. Note that energy is released (negative sign) when the electron falls from the orbital with $n = \infty$ to the one with $n = 1$ and that energy is absorbed (positive sign) when the electron is excited from the orbital with $n = 1$ to the one corresponding to $n = \infty$. Ionization energies are the energies required to remove electrons from atoms, and they are always positive.

1.4 The Photoelectric Effect

In 1887, H. R. Hertz observed that the gap between metal electrodes became a better conductor when ultraviolet light was shined on the apparatus. Soon after, W. Hallwachs observed that a negatively charged zinc surface lost its negative charge when ultraviolet light was shined on it. The negative charges that were lost were identified to be electrons from their behavior in a magnetic field. The phenomenon of an electric current flowing when light was involved became to be known as the *photoelectric effect*.

Studying the photoelectric effect involves an apparatus like that shown schematically in Figure 1.6. An evacuated tube is arranged so that the highly polished metal, such as sodium, potassium, or zinc, to be illuminated is made the cathode. When light shines on the metal plate, electrons flow to the collecting plate (anode), and the ammeter, A, indicates the amount of current flowing. Several observations can be made as the frequency and intensity of the light are varied:

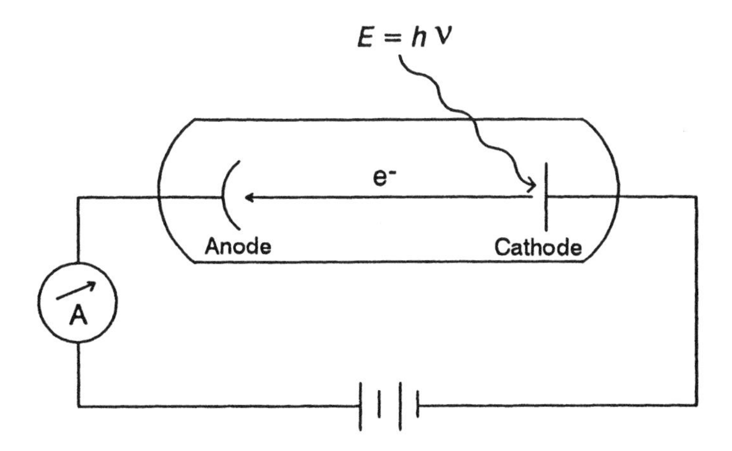

Figure 1.6 ▸ Experimental arrangement for studying the photoelectric effect.

1. The light must have some minimum or threshold frequency, ν_0, in order for the current to flow.

2. Different metals have different threshold frequencies.

3. If the light striking the metal surface has a frequency greater than ν_0, the electrons are ejected with a kinetic energy that increases with the frequency of the light.

4. The number of electrons ejected depends on the intensity of the light but their kinetic energy depends only on the frequency of the light.

An electron traveling toward the collector can be stopped if a negative voltage is applied to the collector. The voltage required (known as the stopping potential, V) to stop the motion of the electrons (causing the current to cease) depends on the frequency of the light that caused the electrons to be ejected. In fact, it is the electrostatic energy of the repulsion between an electron and the collector that exactly equals the kinetic energy of the electron. Therefore, we can equate the two energies by the equation

$$Ve = \frac{1}{2}mv^2. \tag{1.31}$$

An understanding of the photoelectric effect was provided in 1905 by Albert Einstein. Einstein based his analysis on the relationship between the energy of light and its frequency that was established in 1900 by Planck. It was assumed that the light behaved as a collection of particles (called *photons*) and the energy of a particle of light was totally absorbed by the collision

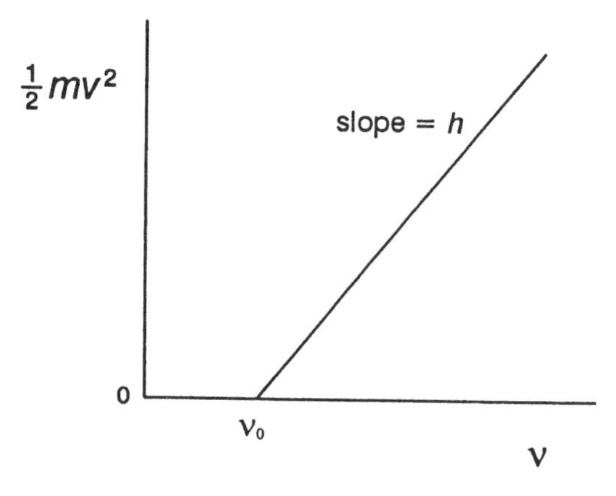

Figure 1.7 ▶ The relationship between the kinetic energy of the ejected electrons and the frequency.

with an electron on the metal surface. Electrons are bound to the surface of a metal with an energy called the *work function*, w, which is different for each type of metal. When the electron is ejected from the surface of the metal, it will have a kinetic energy that is the difference between the energy of the incident photon and the work function of the metal. Therefore, we can write

$$\frac{1}{2}m\nu^2 = h\nu - w. \tag{1.32}$$

It can be seen that this is the equation of a straight line when the kinetic energy of the electron is plotted against the frequency of the light. By varying the frequency of the light and determining the kinetic energy of the electrons (from the stopping potential) a graph like that shown in Figure 1.7 can be prepared to show the relationship. The intercept is ν_0, the threshold frequency, and the slope is Planck's constant, h. One of the significant points in the interpretation of the photoelectric effect is that light is considered to be particulate in nature. In other experiments, such as the diffraction experiment of T. Young, it was necessary to assume that light behaved as a wave. Many photovoltaic devices in common use today (light meters, optical counters, etc.) are based on the photoelectric effect.

1.5 Particle–Wave Duality

Because light behaved as both waves (diffraction, as proved by Young in 1803) and particles (the photoelectric effect shown by Einstein in 1905), the nature of light was debated for many years. Of course, light has characteristics of both a wave and a particle, the so-called *particle–wave duality*. In 1924, Louis de Broglie, a young French doctoral student, investigated some of the consequences of relativity theory. For electromagnetic radiation,

$$E = h\nu = \frac{hc}{\lambda}, \tag{1.33}$$

where c, ν, and λ are the velocity, frequency, and wavelength, respectively, for the radiation. The photon also has an energy given by the relationship from relativity theory,

$$E = mc^2. \tag{1.34}$$

A particular photon has only one energy, so

$$mc^2 = \frac{hc}{\lambda}, \tag{1.35}$$

which can be written as

$$\lambda = \frac{h}{mc}. \tag{1.36}$$

This does not mean that light has a mass, but because mass and energy can be interconverted, it has an energy that is equivalent to some mass. The quantity represented as mass times velocity is the momentum, so Eq. (1.36) predicts a wavelength that is Planck's constant divided by the momentum.

De Broglie reasoned that if a particle had a wave character, the wavelength would be given by

$$\lambda = \frac{h}{mv}, \tag{1.37}$$

where the velocity is written as v rather than c because the particle will not be traveling at the speed of light. This was verified in 1927 by C. J. Davisson and L. H. Germer working at Bell Laboratories in Murray Hill, New Jersey. In their experiment, an electron beam was directed at a metal crystal and a diffraction pattern was observed. Since diffraction is a property of waves, it was concluded that the moving electrons were behaving as waves. The reason for using a metal crystal is that in order to observe a diffraction pattern, the waves must pass through openings about the same size as the wavelength, and that distance corresponds to the distance separating atoms in a metal.

The de Broglie wavelength of moving particles (electrons particularly) has been verified experimentally. This is, of course, important, but the real value is that electron diffraction has now become a standard technique for determining molecular structure.

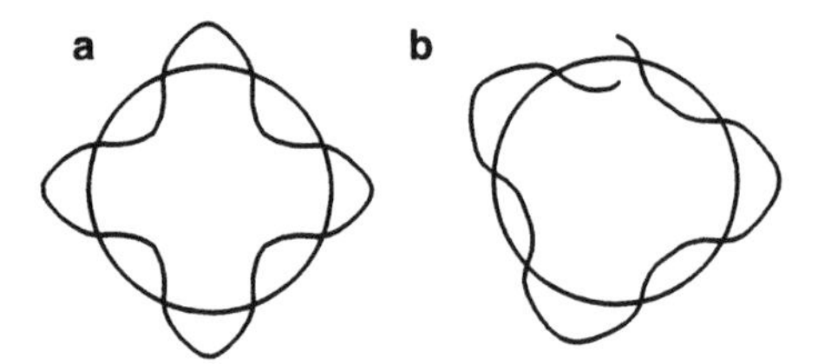

Figure 1.8 ▶ (a) An allowed orbit containing a whole number of wavelengths and (b) an unstable orbit.

In developing a model for the hydrogen atom, Bohr had to assume that the stable orbits were those where

$$mvr = n\frac{h}{2\pi}. \tag{1.38}$$

Because de Broglie showed that the moving electron should be considered as a wave, that wave will be a stable one only if the wave joins smoothly on itself. This means that the circular orbit must contain a whole number of wavelengths (see Figure 1.8). The circumference of a circle in terms of the radius, r, is $C = 2\pi r$. Therefore, a whole number of wavelengths, $n\lambda$, must be equal to the circumference:

$$2\pi r = n\lambda. \tag{1.39}$$

However, the de Broglie wavelength, λ, is given by

$$\lambda = \frac{h}{mv}. \tag{1.40}$$

Therefore,

$$2\pi r = n\left(\frac{h}{mv}\right), \tag{1.41}$$

which can be rearranged to give

$$mvr = n\left(\frac{h}{2\pi}\right), \tag{1.42}$$

which is exactly the same as the equation that Bohr assumed in order to predict which orbits were stable! We now see the connection between the wave character of a particle and the Bohr model. Only two years later, Erwin Schrödinger used the model of a standing wave to represent the electron in the hydrogen atom and solved the resulting wave equation. We will describe this monumental event in science later. While the Bohr model explained the spectral properties of the hydrogen atom, it did not do so for any other atoms. However, He^+, Li^{2+}, and similar species containing one electron could be treated by the same model by making use of the appropriate nuclear charge. Also, the model considered the atom almost like a mechanical device, but since the atom did not radiate energy continuously, it violated laws of electricity and magnetism.

1.6 The Heisenberg Uncertainty Principle

A serious problem with the Bohr model stems from the fact that it is impossible to know simultaneously the position and momentum (or energy) of a particle. A rationale for this can be given as follows. Suppose you observe a ship and determine its position. The visible light waves have a wavelength of about 4×10^{-5} to 8×10^{-5} cm (4×10^{-7} to 8×10^{-7} m) and very low energy. The light strikes the ship and is reflected to your eyes, the detector. Because of the very low energy of the light, the ship, weighing several thousand tons, does not move as a result of light striking it. Now, suppose you wish to "see" a very small particle of perhaps 10^{-8} cm (10^{-10} m) diameter. In order to locate the particle you must use "light" having a wavelength about the same length as the size of the particle. Radiation of 10^{-8} cm (very short) wavelength has very high energy since

$$E = \frac{hc}{\lambda}. \tag{1.43}$$

Therefore, in the process of locating (observing) the particle with high-energy radiation, we have changed its momentum and energy. Therefore, it is impossible to determine both the position and the momentum simultaneously to greater accuracy than some fundamental quantity. That quantity is h and the relationship between the uncertainty in position (distance) and that in momentum (mass $\times$ distance/time) is

$$\Delta x \cdot \Delta (mv) \geq h. \tag{1.44}$$

This relationship, which is one form of the Heisenberg uncertainty principle, indicates that h is the fundamental quantum of action. We can see that this equation is dimensionally correct since the uncertainty in position multiplied by the uncertainty in momentum has the dimensions of

$$\text{distance} \times \left(\text{mass} \times \frac{\text{distance}}{\text{time}}\right).$$

In cgs units,

$$\text{cm} \times \left(\text{g} \times \frac{\text{cm}}{\text{s}}\right) = \text{erg s},$$

and the units of erg s match the units on h.

If we use uncertainty in time in seconds and uncertainty in energy is erg s,

$$\Delta t \cdot \Delta E \geq h, \tag{1.45}$$

and this equation is also dimensionally correct. Therefore, an equation of this form can be written between any two variables that reduce to erg s or g cm^2/s.

It is implied by the Bohr model that we can know the details of the orbital motion of the electron and its energy at the same time. Having now shown that is not true, we will direct our attention to the wave model of the hydrogen atom.

References for Further Reading

- Brown, T. L., LeMay, H. E., Bursten, B., and Burdge, J. R. (2003). *Chemistry, the Central Science*, 9th ed. Prentice Hall, Englewood Cliffs, NJ. One of the standard texts in general chemistry.
- Krane, K. (1995). *Modern Physics*, 2nd ed. Wiley, New York. A good, readable book on many topics related to atomic physics.
- Pauling, L. (1988). *General Chemistry*. Dover, New York. A widely available, inexpensive source of an enormous amount of information written by a legend in science.
- Porile, N. T. (1993). *Modern University Chemistry*, 2nd ed. McGraw–Hill, New York. A general chemistry book that is somewhat more advanced than most. Chapter 5 gives an excellent introduction to quantum phenomena related to atomic structure.
- Serway, R. E. (2000). *Physics for Scientists and Engineers*, 5th ed. Saunders (Thomson Learning), Philadelphia. Excellent coverage of all areas of physics.

Problems

1. A shortwave radio station in Lake Okeechobee, Florida, broadcasts on a frequency of 6.065 megahertz (MHz). What is the wavelength of the radio waves?

2. What would be the de Broglie wavelength of an electron moving at 2.00% of the speed of light?

3. An electron in the ground state of a hydrogen atom is struck by an X-ray photon having a wavelength of 50.0 nm. A scattered photon having a wavelength of 200 nm is observed after the collision. What will be the velocity and de Broglie wavelength of the ejected electron?

4. The *work function* of a metal is the energy required to remove an electron from the metal. What wavelength of light will just eject an electron from a metal that has a work function of 2.60 electron volts (eV) ($1\ \text{eV} = 1.60 \times 10^{-12}$ erg)?

5. For Be^{3+}, calculate the wavelength of the photons emitted as the electron falls from $n = 3$ to $n = 2$ and from $n = 4$ to $n = 3$.

6. Lithium compounds containing Li^{+} ions impart a red color to a flame due to light emitted that has a wavelength of 670.8 nm.

(a) What is the frequency of this spectral line?

(b) What is the wave number for the radiation?

(c) In kcal/mol, what energy is associated with this spectral line?

7. The ionization potential for a certain atom is 350 kJ/mol. If the electron is in the first excited state, the ionization potential is 105 kJ/mol. If the atom undergoes deexcitation, what would be the wavelength of the photon emitted?

8. The work function for barium is 2.48 eV. If light of 400 nm is shined on a barium cathode, what is the maximum velocity of the ejected electrons?

9. Creation of matter from electromagnetic radiation can occur if the radiation has sufficient energy (pair production). What is the minimum energy of a photon that could produce an electron–positron (a positive electron) pair?

10. For a proton and an electron having kinetic energies of 2.0 eV ($1\ \text{eV} = 1.6 \times 10^{-12}$ erg), what would be the ratio of the de Broglie wavelengths?

11. Neutrons having energies equivalent to the kinetic energy of gaseous molecules at room temperature (kT) are called *thermal neutrons.* What is the wavelength of a thermal neutron at 27°C?

12. Suppose an electron remains in an excited state of an atom for 10^{-8} s. What would be the minimum uncertainty of the energy of the photon emitted as the electron falls to the ground state? What uncertainty in the wavelength of the photon does this correspond to?

13. The ionization potential for electrons not used in bonding in the water molecule is 12.6 eV. If a single X-ray photon having a wavelength of 0.300 nm passes through water and is completely absorbed by ionizing water molecules, how many molecules could the photon ionize? Explain why in practice this would not be likely to occur.

14. For a gas, the root-mean-square velocity is given by $v = (3RT/M)^{1/2}$, where M is the molecular weight, T is the temperature in Kelvin, and R is the molar gas constant. Derive an expression for the de Broglie wavelength of gaseous molecules at a temperature T. Use the expression to determine the wavelength of moving helium atoms at a temperature of 300 K.

15. Repeat the procedure of Problem 14 except use the average velocity for gaseous molecules, $v_a = (8RT/\pi M)^{1/2}$. After deriving the relationship, determine the de Broglie wavelength for hydrogen molecules at 400 K.

16. Radon-212 emits an alpha particle (helium nucleus) having an energy of 6.26 MeV. Determine the wavelength of the alpha particle. To a good approximation, the radius of a nucleus (in centimeters) can be expressed as $R = r_0 A^{1/3}$, where r_0 is a constant with a value of 1.3×10^{-13} and A is the mass number. Compare the wavelength of the alpha particle emitted from ^{212}Rn with the diameter of the nucleus.

17. Show from the Bohr model that the ratio of the kinetic energy to the potential energy is $-\frac{1}{2}$.

18. From the relationships used in the Bohr model, show that the velocity of the electron in the first Bohr orbit is $1/137$ of the velocity of light.

19. Show that the difference in energy between any two spectral lines in the hydrogen atom is the energy corresponding to a third spectral line. This phenomenon is known as the Ritz Principle.

20. One form of the Heisenberg Uncertainty Principle is $\Delta E \times \Delta t \geq h$ where ΔE is the uncertainty in energy and Δt is the uncertainty in the time. If it requires 10^{-8} s for an electron to fall from a higher orbital to a lower one, what will be the width of the spectral line emitted?

▶ Chapter 2

The Quantum Mechanical Way of Doing Things

There are several areas of chemistry that require a knowledge of quantum mechanics for their explanation and understanding. Therefore quantum mechanics at an elementary level is covered in several physics and chemistry courses taken by undergraduates. We will introduce some of the procedures and terminology of quantum mechanics by stating the postulates of quantum mechanics and showing some of their applications. The coverage here is meant to be a very brief introduction to the field and in no way is adequate for a complete understanding of this important topic. For more complete coverage of quantum mechanics and its applications, see the references listed at the end of this chapter. We begin our exploration of quantum mechanics with postulates and their meanings.

2.1 The Postulates

POSTULATE I. For any possible state of a system, there is a function, Ψ, of the coordinates of the parts of the system and time that completely describes the system.

For a single particle described by Cartesian coordinates, we can write

$$\Psi = \Psi(x, y, z, t). \tag{2.1}$$

For two particles, the coordinates of each particle must be specified so that

$$\Psi = \Psi(x_1,\ y_1,\ z_1,\ x_2,\ y_2,\ z_2,\ t). \tag{2.2}$$

For a general system, we can use generalized coordinates, q_i,

$$\Psi = \Psi\,(q_i,\ t)\,. \tag{2.3}$$

Since the model is that of a wave, Ψ is called a *wave function*. The state of the system that i describes is called the *quantum state*.

The wave function squared, Ψ^2, is proportional to probability. Since Ψ may be complex, we are interested in $\Psi^*\Psi$, where Ψ^* is the *complex conjugate* of Ψ. The complex conjugate is the same function with i replaced by $-i$, where $i = \sqrt{-1}$. For example, if we square the function $(x + ib)$ we obtain

$$(x + ib)\,(x + ib) = x^2 + 2ibx + i^2b^2 = x^2 + 2ibx - b^2, \tag{2.4}$$

and the resulting function is still complex. If we multiply $(x + ib)$ by its complex conjugate $(x - ib)$, we obtain

$$(x + ib)\,(x - ib) = x^2 - i^2b^2 = x^2 + b^2, \tag{2.5}$$

which is a real function.

The quantity $\Psi^*\Psi\, d\tau$ is proportional to the probability of finding the particles of the system in the volume element, $d\tau = dx\, dy\, dz$. We require that the total probability be unity (1) so that the particle must be *somewhere.* That is,

$$\int_{\text{all space}} \Psi^*\Psi\, d\tau = 1. \tag{2.6}$$

If this condition is met, then Ψ is *normalized*. In addition, Ψ must be *finite, single valued,* and *continuous*. These conditions describe a "well-behaved" wave function. The reasons for these requirements are as follows:

FINITE A probability of unity (1.00) denotes a "sure thing." A probability of 0 means that a certain event cannot happen. Therefore, probability varies from 0 to unity. If Ψ were infinite, the probability could be greater than 1.

SINGLE VALUED In a given region of space, there is only one probability of finding a particle. Similarly, in a hydrogen atom, there is a single probability of finding the electron at some specified distance from the nucleus. There are not two different probabilities of finding the electron at some given distance from the nucleus.

CONTINUOUS If there is a certain probability of finding an electron at a given distance from the nucleus in a hydrogen atom, there will be a slightly different probability if the distance is changed slightly. The probability does not suddenly double if the distance is changed by 0.01%. The probability function does not have discontinuities so the wavefunction must be continuous.

If two functions ϕ_1 and ϕ_2 have the property that

$$\int \phi_1^* \phi_2 \, d\tau = 0 \quad \text{or} \quad \int \phi_1 \phi_2^* \, d\tau = 0 \tag{2.7}$$

they are said to be *orthogonal*. Whether the integral vanishes may depend on the limits of integration, and hence one speaks of *orthogonality* within a certain interval. Therefore, the limits of integration must be clear. In the previous case, the integration is carried out over the possible range of the coordinates used in $d\tau$. If the coordinates are x, y, and z, the limits are from $-\infty$ to $+\infty$ for each variable. If the coordinates are r, θ, and ϕ, the limits of integration are 0 to ∞, 0 to π, and 0 to 2π, respectively. We will have more to say about orthogonal wave functions in later chapters.

2.2 The Wave Equation

It was shown in 1924 by de Broglie that a moving particle has a wave character. That idea was demonstrated in 1927 by Davisson and Germer when an electron beam was diffracted by a nickel crystal. Even before that experimental verification, Erwin Schrödinger adapted the wave model to the problem of the hydrogen atom. In that case, the model needs to describe a three-dimensional wave. Classical physics had dealt with such models in a problem known as the flooded planet problem. This model considers the waveforms that would result from a disturbance of a sphere that is covered with water. The classical three-dimensional wave equation is

$$\frac{\partial^2 \phi}{\partial x^2} + \frac{\partial^2 \phi}{\partial y^2} + \frac{\partial^2 \phi}{\partial z^2} = \frac{1}{v^2} \frac{\partial^2 \phi}{\partial t^2}, \tag{2.8}$$

where ϕ is the amplitude function and v is the phase velocity of the wave. For harmonic motion (like a sine wave)

$$\frac{\partial^2 \phi}{\partial t^2} = -4\pi^2 \nu^2 \phi, \tag{2.9}$$

where ν is the frequency. The de Broglie relationship

$$\lambda = \frac{h}{mv} \tag{2.10}$$

and the relationship between the frequency and the wavelength of a transverse wave,

$$\text{distance} \times \text{time}^{-1} = v\,(\text{distance/time})\,,$$

allow us to write

$$\nu^2 = m^2 v^4 / h^2. \tag{2.11}$$

Therefore, we find that

$$\frac{\partial^2 \phi}{\partial t^2} = -4\pi^2 \left(\frac{m^2 v^4}{h^2}\right) \phi. \tag{2.12}$$

Substituting this in the general equation above, we obtain

$$\frac{\partial^2 \phi}{\partial x^2} + \frac{\partial^2 \phi}{\partial y^2} + \frac{\partial^2 \phi}{\partial z^2} = \left(\frac{1}{\nu^2}\right)\left(\frac{-4\pi^2 v^4 m^2}{h^2}\right) \phi. \tag{2.13}$$

Now let E be the total energy, T be the kinetic energy, and V be the potential energy. Then

$$T = \frac{mv^2}{2} = E - V. \tag{2.14}$$

Substituting this into Eq. (2.13) gives

$$\frac{\partial^2 \phi}{\partial x^2} + \frac{\partial^2 \phi}{\partial y^2} + \frac{\partial^2 \phi}{\partial z^2} = \left(\frac{-4\pi^2 m^2}{h^2}\right)\left[\frac{2(E - V)}{m}\right] \phi. \tag{2.15}$$

This equation, which describes a three-dimensional wave, can be rearranged and written in the form

$$\frac{\partial^2 \phi}{\partial x^2} + \frac{\partial^2 \phi}{\partial y^2} + \frac{\partial^2 \phi}{\partial z^2} + \left(\frac{8\pi^2 m}{h^2}\right)(E - V)\phi = 0. \tag{2.16}$$

This is one form of the *Schrödinger wave equation*. Solutions are the *wave functions*, and solving wave equations involves that branch of science known as *wave mechanics* (quantum mechanics). The preceding presentation is not a derivation of the Schrödinger wave equation. It is an adaptation of a classical wave equation to a different system by use of the de Broglie hypothesis. It is interesting to note that Schrödinger's treatment of the hydrogen atom started with an equation that was already known. As we shall see in Chapter 4, the solution of the wave equation for hydrogen makes use of mathematical techniques that were already in existence at the time. While Schrödinger's work was phenomenal, it was not carried out without the understanding of what had been done before. In fact, his incorporation of de Broglie's hypothesis came only a couple of years after the idea became known.

2.3 Operators

POSTULATE II. For every dynamical variable (classical observable) there is a corresponding operator.

This postulate provides the connection between quantities that are classical observables and the quantum mechanical techniques for doing things. Dynamical variables are such quantities as energy, momentum, angular momentum, and position coordinates. In quantum mechanics, these are replaced by *operators*. Operators are symbols that indicate that some mathematical operation is to be performed. Such symbols include $(\,)^2, d/dx$, and $\int$. Coordinates are the same in operator and classical forms. For example, the coordinate x is simply used in operator form as x. This will be illustrated later. Other classical observables are replaced by operators shown in Table 2.1. Also, as we shall see, some other operators can be

TABLE 2.1 ▶ Some Operators Frequently Used in Quantum Mechanics

Quantity	Symbol used	Operator form
Coordinates	x, y, z, r	x, y, z, r
Momentum		
x	p_x	$\frac{\hbar}{i}\frac{\partial}{\partial x}$
y	p_y	$\frac{\hbar}{i}\frac{\partial}{\partial y}$
z	p_z	$\frac{\hbar}{i}\frac{\partial}{\partial z}$
Kinetic energy	$\frac{p^2}{2m}$	$-\frac{\hbar^2}{2m}\left(\frac{\partial^2}{\partial x^2}+\frac{\partial^2}{\partial y^2}+\frac{\partial^2}{\partial z^2}\right)$
Kinetic energy	T	$-\frac{\hbar}{i}\frac{\partial}{\partial t}$
Potential energy	V	$V(q_i)$
Angular momentum	L_z (Cartesian)	$\frac{\hbar}{i}\left(x\frac{\partial}{\partial y}-y\frac{\partial}{\partial x}\right)$
	L_z (polar)	$\frac{\hbar}{i}\frac{\partial}{\partial \phi}$

formed by combining those in the table. For example, since the kinetic energy is $mv^2/2$, it can be written in terms of the momentum, p, as $p^2/2m$.

The operators that are important in quantum mechanics have two important characteristics. First, the operators are linear, which means that

$$\alpha\left(\phi_1+\phi_2\right)=\alpha\phi_1+\alpha\phi_2, \tag{2.17}$$

where α is the operator and ϕ_1 and ϕ_2 are the functions being operated on. Also, if C is a constant, then

$$\alpha\left(C\phi\right)=C\left(\alpha\phi\right). \tag{2.18}$$

The linear character of the operator is related to the superposition of states and waves reinforcing each other in the process.

The second property of the operators that we will see in quantum mechanics is that they are Hermitian. If we consider two functions ϕ_1 and ϕ_2, the operator α is Hermitian if

$$\int \phi_1^*\alpha\phi_2\,d\tau=\int \phi_2\alpha^*\phi_1^*\,d\tau. \tag{2.19}$$

This requirement is necessary to ensure that the calculated quantities are real. We will have several opportunities to observe these types of behavior in the operators we will use.

2.4 Eigenvalues

POSTULATE III. The permissible values that a dynamical variable may have are those given by $\alpha\phi=a\phi$, where ϕ is the *eigenfunction* of the operator α that corresponds to the observable whose permissible values are a.

This postulate can be stated in the form of an equation as

$$\underset{\text{operator}}{\alpha}\quad\underset{\text{wave function}}{\phi}\quad=\quad\underset{\text{constant (eigenvalue)}}{a}\quad\underset{\text{wave function}}{\phi}. \tag{2.20}$$

If performing the operation on the wave function yields the original function multiplied by a constant, then ϕ is an eigenfunction of the operator α. This can be illustrated by letting $\phi=\mathrm{e}^{2x}$ and the operator $\alpha=d/dx$. Then operating on the function with the operator gives

$$d\phi/dx=2\,\mathrm{e}^{2x}=\text{constant}\cdot\mathrm{e}^{2x} \tag{2.21}$$

Therefore, e^{2x} is an eigenfunction of the operator α with an eigenvalue of 2.

If we let $\phi = e^{2x}$ and the operator be $(\ )^2$, we obtain

$$\left(e^{2x}\right)^2 = e^{4x},$$

which is not a constant times the original function. Therefore, e^{2x} is not an eigenfunction of the operator $(\ )^2$.

If we use the operator for the z component of angular momentum,

$$\hat{L}_z = \frac{\hbar}{i}\frac{\partial}{\partial\phi}, \tag{2.22}$$

(where $\hbar = h/2\pi$) operating on the function $e^{in\phi}$ (where n is a constant), we obtain

$$\frac{\hbar}{i}\frac{\partial}{\partial\phi}\left(e^{in\phi}\right) = in\frac{\hbar}{i}e^{in\phi} = n\hbar \cdot e^{in\phi}, \tag{2.23}$$

which is a constant ($n\hbar$) times the original function. Therefore, the eigenvalue is $n\hbar$.

For a given system, there may be various possible values of a parameter we wish to calculate. Since most properties (such as the distance from the nucleus to an electron) may vary, we desire to determine an *average* or *expectation* value. Using the operator equation $\alpha\phi = a\phi$, where ϕ is some function, we multiply both sides of the equation by ϕ^*:

$$\phi^*\alpha\phi = \phi^* a\phi. \tag{2.24}$$

Note, however, that $\phi^*\alpha\phi$ is not necessarily the same as $\phi\alpha\phi^*$. To obtain the sum of the probability over all space, we write this in the form of the integral equation,

$$\int_{\text{all space}} \phi^*\alpha\phi\, d\tau = \int_{\text{all space}} \phi^* a\phi\, d\tau \tag{2.25}$$

However, a is a constant and is not affected by the order of operations. Removing it from the integral and solving for a yields

$$a = \frac{\int \phi^*\alpha\phi\, d\tau}{\int \phi^*\phi\, d\tau}. \tag{2.26}$$

Remember that since α is an operator, $\phi^*\alpha\phi$ is not necessarily the same as $\alpha\phi\phi^*$, so that the order $\phi^*\alpha\phi$ must be preserved and α cannot be removed from the integral.

If ϕ is normalized, then by definition $\int \phi^* \phi \, d\tau = 1$ and

$$\bar{a} = \langle a \rangle = \int \phi^* \alpha \phi \, d\tau, \tag{2.27}$$

where $\bar{a}$ and $\langle a \rangle$ are the usual ways of indicating the average or expectation value. If the wave function is known, then theoretically an expectation or average value can be calculated for a given parameter by using the appropriate operator and the procedure shown above.

Consider the following example, which illustrates the application of these ideas: Suppose we want to calculate the radius of the hydrogen atom in the $1s$ state. The normalized wave function is

$$\psi_{1s} = \left(\frac{1}{\sqrt{\pi}}\right)\left(\frac{1}{a_0}\right)^{3/2} \mathrm{e}^{-r/a_0} = \psi_{1s}^*, \tag{2.28}$$

where a_0 is the first Bohr radius. The equation becomes

$$\langle r \rangle = \int \psi^* \text{ (operator) } \psi \, d\tau. \tag{2.29}$$

The operator here is just r since position coordinates have the same form in operator and classical form. In polar coordinates, the volume element $d\tau = r^2 \sin\theta \, dr \, d\theta \, d\phi$. Therefore, the problem becomes

$$\begin{aligned} \langle r \rangle = & \int_0^\infty \int_0^\pi \int_0^{2\pi} \frac{1}{\sqrt{\pi}} \left(\frac{1}{a_0}\right)^{3/2} \mathrm{e}^{-r/a_0} (r) \frac{1}{\sqrt{\pi}} \left(\frac{1}{\sqrt{a_0}}\right)^{3/2} \\ & \times \mathrm{e}^{-r/a_0} r^2 \sin\theta \, dr \, d\theta \, d\phi. \end{aligned} \tag{2.30}$$

Although this may look rather formidable, it simplifies greatly since the operator r becomes a multiplier and the functions of r can be multiplied. The result is

$$\langle r \rangle = \int_0^\infty \int_0^\pi \int_0^{2\pi} \frac{1}{\pi a_0^3} r^3 \mathrm{e}^{-r/a_0} \sin\theta \, dr \, d\theta \, d\phi. \tag{2.31}$$

Using the technique from calculus that enables us to separate multiple integrals of the type

$$\int f(x) \, g(y) \, dx \, dy = \int f(x) \, dx \int g(y) \, dy, \tag{2.32}$$

we can write Eq. (2.31) as

$$\langle r \rangle = \frac{1}{\pi a_0^3} \int_0^\infty r^3 \mathrm{e}^{-2r/a_0} \, dr \int_0^\pi \int_0^{2\pi} \sin\theta \, d\theta \, d\phi. \tag{2.33}$$

It is easily verified that

$$\int_0^{\pi}\int_0^{2\pi} \sin\theta\, d\theta\, d\phi = 4\pi, \tag{2.34}$$

and the exponential integral is a commonly occurring one in quantum mechanics. It is easily evaluated using the formula

$$\int_0^{\infty} x^n \mathrm{e}^{-bx}\, dx = \frac{n!}{b^{n+1}}. \tag{2.35}$$

In this case, $n = 3$ and $b = 2/a_0$. Therefore,

$$\int_0^{\infty} r^3 \mathrm{e}^{-2r/a_0}\, dr = \frac{3!}{(2/a_0)^4}, \tag{2.36}$$

so that

$$\langle r \rangle = \frac{4\pi}{\pi a_0^3}\frac{3!}{(2/a_0)^4} = \frac{3}{2}a_0. \tag{2.37}$$

Thus, $\langle r \rangle_{1s} = (3/2)a_0 \left(a_0 = 0.529\ \text{Å}\right)$.

The average distance of the electron from the nucleus in the $1s$ state of hydrogen is 3/2 the radius of the first Bohr radius. However, the most

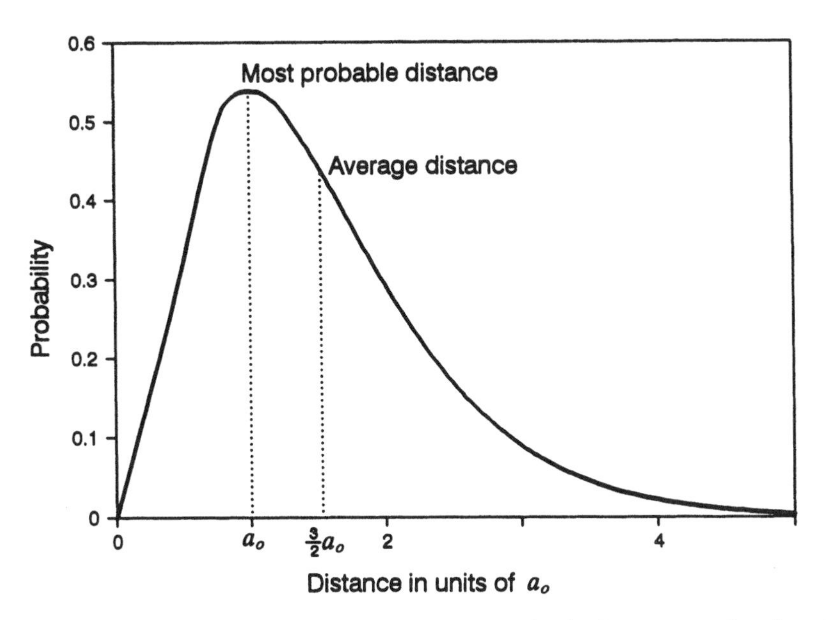

Figure 2.1 ▶ The probability of finding the electron in the $1s$ state as a function of distance from the nucleus.

probable distance is the same as the radius of the first Bohr orbit. *Average* and *most probable* are not the same. The reason for this is that the probability distribution is not symmetric, as is shown in Figure 2.1.

2.5 Wave Functions

POSTULATE IV. The state function, Ψ, is given as a solution of

$$\hat{H}\Psi = E\Psi, \tag{2.38}$$

where $\hat{H}$ is the operator for total energy, the *Hamiltonian operator.*

This postulate provides a starting point for formulating a problem in quantum mechanical terms because we usually seek to determine a wave function to describe the system being studied. The *Hamiltonian function* in classical physics is the total energy, $T + V$, where T is the translational (kinetic) energy and V is the potential energy. In operator form,

$$\hat{H} = \hat{T} + \hat{V}, \tag{2.39}$$

where $\hat{T}$ is the operator for kinetic energy and $\hat{V}$ is the operator for potential energy. Written in terms of the generalized coordinates, q_i, and time, the starting equation becomes

$$\hat{H}\Psi(q_i, t) = -\frac{\hbar}{i}\frac{\partial \Psi(q_i, t)}{\partial t}. \tag{2.40}$$

The kinetic energy can be written in terms of the momentum as

$$T = mv^2/2 = p^2/2m. \tag{2.41}$$

We now write

$$T = \frac{p_x^2}{2m} + \frac{p_y^2}{2m} + \frac{p_z^2}{2m}. \tag{2.42}$$

Putting this in operator form, we make use of the momentum operators given earlier. Then

$$\hat{T} = \frac{1}{2m}\left(\frac{\hbar}{i}\frac{\partial}{\partial x}\right)^2 + \frac{1}{2m}\left(\frac{\hbar}{i}\frac{\partial}{\partial y}\right)^2 + \frac{1}{2m}\left(\frac{\hbar}{i}\frac{\partial}{\partial z}\right)^2. \tag{2.43}$$

However, we can write the square of each momentum operator as

$$\left(\frac{\hbar}{i}\frac{\partial}{\partial x}\right)^2 = \left(\frac{\hbar}{i}\frac{\partial}{\partial x}\right)\left(\frac{\hbar}{i}\frac{\partial}{\partial x}\right) = \left(\frac{\hbar^2}{i^2}\frac{\partial^2}{\partial x^2}\right) = -\hbar^2\frac{\partial^2}{\partial x^2} \tag{2.44}$$

so that

$$\hat{T} = -\frac{\hbar^2}{2m}\left(\frac{\partial^2}{\partial x^2} + \frac{\partial^2}{\partial y^2} + \frac{\partial^2}{\partial z^2}\right) = -\frac{\hbar^2}{2m}\nabla^2, \tag{2.45}$$

where ∇^2 is the *Laplacian operator* or simply the Laplacian. The general form of the potential energy can be written as

$$V = V(q_i, t) \tag{2.46}$$

so that the operator equation becomes

$$\left(-\frac{\hbar^2}{2m}\nabla^2 + V(q_i, t)\right)\Psi(q_i, t) = -\frac{\hbar}{i}\frac{\Psi(q_i, t)}{\partial t}. \tag{2.47}$$

This equation is known as Schrödinger's *time-dependent* equation or Schrödinger's *second* equation.

In most problems, the classical observables have values that do not change with time, or at least their average values do not change with time. Therefore, for most cases, it would be advantageous to simplify the problem by removal of the dependence on the time. The separation of variables technique is now applied to see whether the time dependence can be separated. The separation of variables as a technique in differential equations is discussed in Chapters 3 and 4, but we will use it here with very little explanation. First, it is assumed that $\Psi(q_i, t)$ is the product of two functions, one a function that contains only q_i and another that contains only the time, t. Then we can write

$$\Psi(q_i, t) = \psi(q_i)\,\tau(t). \tag{2.48}$$

Note that Ψ is used to denote the complete state function and the lower case ψ is used to represent the state function with the time dependence removed. Since the problems that we will consider in this book are time-independent ones, ψ will be used throughout. The Hamiltonian can now be written in terms of the two functions ψ and τ as

$$\hat{H}\Psi(q_i, t) = \hat{H}\psi(q_i)\,\tau(t). \tag{2.49}$$

Therefore, since $\psi(q_i)$ is not a function of t, Eq. (2.49) can be written as

$$\hat{H}\psi(q_i)\,\tau(t) = -\frac{\hbar}{i}\frac{\partial}{\partial t}\psi(q_i)\,\tau(t) = -\frac{\hbar}{i}\psi(q_i)\frac{\partial\tau(t)}{\partial t}. \tag{2.50}$$

Dividing Eq. (2.50) by the product $\psi(q_i)\,\tau(t)$,

$$\frac{\hat{H}\psi(q_i)\,\tau(t)}{\psi(q_i)\,\tau(t)} = \frac{-(\hbar/i)\,\psi(q_i)\,[\partial\tau(t)/\partial t]}{\psi(q_i)\,\tau(t)}, \tag{2.51}$$

we obtain

$$\frac{1}{\psi\,(q_i)}\hat{H}\psi\,(q_i) = -\frac{\hbar}{i}\frac{1}{\tau\,(t)}\frac{\partial\tau\,(t)}{\partial t}. \tag{2.52}$$

Note that the factor $\psi\,(q_i)$ does not cancel on the left-hand side of Eq. (2.52) since $\hat{H}\psi\,(q_i)$ does not represent $\hat{H}$ times $\psi\,(q_i)$ but rather $\hat{H}$ *operating on* $\psi\,(q_i)$. The left-hand side is a function of q_i and the right-hand side is a function of t, so each can be considered as a constant with respect to changes in the values of the other variable. Both sides can be set equal to some new parameter, W, so that

$$\frac{1}{\psi\,(q_i)}\hat{H}\psi\,(q_i) = W \quad \text{and} \quad -\frac{\hbar}{i\tau\,(t)}\frac{\partial\tau\,(t)}{\partial t} = W. \tag{2.53}$$

From the first of these equations, we obtain

$$\hat{H}\psi\,(q_i) = W\psi\,(q_i)\,, \tag{2.54}$$

and from the second we obtain

$$\frac{d\tau\,(t)}{dt} = -\frac{i}{\hbar}W\tau\,(t)\,. \tag{2.55}$$

The differential equation involving the time can be solved readily to give

$$\tau\,(t) = \mathrm{e}^{-(i/\hbar)Wt}. \tag{2.56}$$

Substituting this result into Eq. (2.48), we find that the total state function, Ψ, is

$$\Psi\,(q_i, t) = \psi\,(q_i)\,\mathrm{e}^{-(i/\hbar)Wt}. \tag{2.57}$$

Therefore, Eq. (2.50) can be written as

$$\mathrm{e}^{-(i/\hbar)Wt}\hat{H}\psi\,(q_i) = -\frac{\hbar}{i}\frac{i}{\hbar}W\psi\,(q_i)\,\mathrm{e}^{-(i/\hbar)Wt} \tag{2.58}$$

or

$$\mathrm{e}^{-(i/\hbar)Wt}\hat{H}\psi\,(q_i) = W\psi\,(q_i)\,\mathrm{e}^{-(i/\hbar)Wt}. \tag{2.59}$$

The factor $\mathrm{e}^{-(i/\hbar)Wt}$ can be dropped from both sides of Eq. (2.59), which results in

$$\hat{H}\psi\,(q_i) = W\psi\,(q_i)\,, \tag{2.60}$$

which shows that the time dependence has been separated.

In this equation, neither the Hamiltonian operator nor the wave function is time dependent. It is this form of the equation that will be used to solve problems discussed in this book. Therefore, the time-independent wave function, ψ, will be indicated any time we write $\hat{H}\psi = E\psi$.

For the hydrogen atom, $V = -e^2/r$, which remains unchanged in operator form since e is a constant and r represents a coordinate. Therefore,

$$\hat{H} = -\frac{\hbar^2}{2m}\nabla^2 - \frac{e^2}{r}, \tag{2.61}$$

which gives

$$\hat{H}\psi = E\psi = -\frac{\hbar^2}{2m}\nabla^2\psi - \frac{e^2}{r}\psi \tag{2.62}$$

or

$$\nabla^2\psi + \frac{2m}{\hbar^2}(E - V)\psi = 0. \tag{2.63}$$

This is the Schrödinger wave equation for the hydrogen atom.

Several relatively simple models are capable of being treated by the methods of quantum mechanics. To treat these models, we use the four postulates in a relatively straightforward manner. For any of these models, we begin with

$$\hat{H}\psi = E\psi \tag{2.64}$$

and use the appropriate expressions for the operators corresponding to the potential and kinetic energies. In practice, we will find that there is a rather limited number of potential functions, the most common being a Coulombic (electrostatic) potential.

The quantum mechanical models are presented because they can be applied to several systems of considerable interest. For example, the barrier penetration phenomenon has application as a model for nuclear decay and transition state theory in chemical kinetics. The rigid rotor and harmonic oscillator models are useful as models in rotational and vibrational spectroscopy. The particle in the box model has some utility in treating electrons in metals or conjugated molecules. Given the utility of these models, some familiarity with each of them is essential for all who would understand the application of quantum mechanics to problems of relevance to a wide range of sciences. The next several chapters will deal with the basic models and their applications.

References for Further Reading

- Alberty, R. A., and Silbey, R. J. (1996). *Physical Chemistry*, 2nd ed. Wiley, New York. The chapters on quantum mechanics provide an excellent survey of the field.
- Eyring, H., Walter, J., and Kimball, G. E. (1944). *Quantum Chemistry*. Wiley, New York. One of the two true classics in the applications of quantum mechanics to chemistry.

- Hameka, H. F. (1967). *Introduction to Quantum Theory*. Harper & Row, New York. An older book worth finding that presents a lot of mathematical detail. A good intermediate level book.
- Laidler, K. J., and Meiser, J. H. (1982). *Physical Chemistry*, Chap. 11. Benjamin–Cummings, Menlo Park, CA. A clear introduction to the basic methods of quantum mechanics.
- Pauling, L., and Wilson, E. B. (1935). *Introduction to Quantum Mechanics*. McGraw–Hill, New York. The other classic in quantum mechanics aimed primarily at chemistry. Now available in an inexpensive, widely available reprint from Dover.
- Sherwin, C. W. (1959). *Introduction to Quantum Mechanics*. Holt, Rinehart, & Winston, New York. One of the best overall accounts of quantum mechanics. Highly recommended.

Problems

1. The operator for the z component of angular momentum $\hat{L}_z$ in polar coordinates is $(\hbar/i)\,(\partial/\partial\phi)$. Determine which of the following functions are eigenfunctions of this operator and determine the eigenvalues for those that are [in (b), l is an integer constant]:

(a) $\sin\phi e^{i\phi}$,

(b) $\sin^l\phi e^{il\phi}$, and

(c) $\sin\phi e^{-i\phi}$.

2. Calculate the expectation value for the z component of angular momentum for functions (a) and (b) in Problem 1.

3. Normalize the following functions in the interval 0 to ∞

(a) e^{-5x}, and

(b) e^{-bx} (where b is a constant).

4. Show that the function $(x+iy)/r$ is an eigenfunction of the operator for the z component of angular momentum (see Table 2.1).

5. Show that the 1s wave function for hydrogen,

$$\psi_{1s} = \frac{1}{\sqrt{\pi}}\left(\frac{1}{a_0}\right)^{3/2} e^{-r/a_0},$$

is normalized.

6. Show that $\psi = ae^{-bx}$ (where a and b are constants) is an eigenfunction of the operator d^2/dx^2.

7. Normalize the function $\phi = ae^{-bx}$ in the interval 0 to ∞.

8. Determine whether the function $\phi = \sin\, xe^{ax}$ (where a is a constant) is an eigenfunction of the operators d/dx and d^2/dx^2. If it is, determine any eigenvalue(s).

9. Normalize the function $\psi = \sin(\pi x/L) + i\sin(2\pi x/L)$ in the interval 0 to L. Determine the expectation values for the momentum, p, and the kinetic energy, T.

10. Functions and operators are said to be symmetric if $f(x) = f(-x)$. Determine whether the operator for kinetic energy is symmetric or antisymmetric.

Chapter 3

Particles in Boxes

As one begins the study of quantum mechanics, it is desirable to consider some simple problems that can be solved exactly whether they represent important models of nature or not. Such is the case with the models of particles in boxes. Although they have some applicability to real problems, such models are most useful in illustrating the methods of formulating problems and applying quantum mechanical procedures. As a result, almost every introductory book on quantum mechanics includes discussion on particles in boxes.

3.1 The Particle in a One-Dimensional Box

In this model, we treat the behavior of a particle that is confined to motion in a box. The box is taken to be one dimensional for simplicity although a three-dimensional problem is not much more difficult, and we will take up that problem next. To confine the particle absolutely to the box, we make the walls of the box infinitely high. Otherwise, there is a small but finite probability that the particle can "leak" out of the box by *tunneling*. We shall treat the problem of tunneling through a potential energy barrier later.

The coordinate system for this problem is shown in Figure 3.1. The Hamiltonian, H, is

$$H = T + V = \frac{p^2}{2m} + V, \tag{3.1}$$

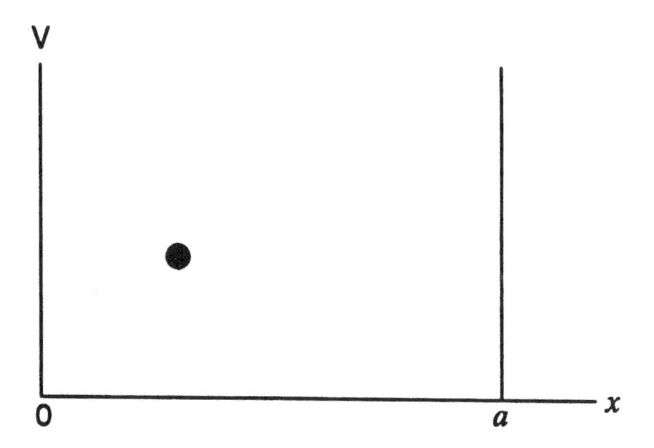

Figure 3.1 ▶ A particle in a one-dimensional box.

where p is the momentum, m is the mass of the particle, and V is the potential energy. Outside the box, $V = \infty$, so

$$H = \frac{p^2}{2m} + \infty, \tag{3.2}$$

the Hamiltonian operator, $\hat{H}$, is

$$\hat{H} = -\frac{\hbar^2}{2m}\frac{d^2}{dx^2} + \infty, \tag{3.3}$$

and the wave equation becomes

$$\hat{H}\psi = E\psi = -\frac{\hbar^2}{2m}\frac{d^2}{dx^2}\psi + \infty\psi. \tag{3.4}$$

Therefore, for the equation to be valid, ψ must be 0 and the probability of finding the particle outside the box is zero.

Inside the box, the potential energy is zero ($V = 0$) so the wave equation can be written as

$$\hat{H}\psi = E\psi \tag{3.5}$$

or

$$-\frac{\hbar^2}{2m}\frac{d^2\psi}{dx^2} = E\psi. \tag{3.6}$$

This equation can be written as

$$\frac{d^2\psi}{dx^2} + k^2\psi = 0, \tag{3.7}$$

where $k^2 = 2mE/\hbar^2$. This is a linear differential equation with constant coefficients, which is a standard type of equation having a solution of the form

$$\psi = A\cos kx + B\sin kx. \tag{3.8}$$

Actually, we could guess a form of the solution in this case since Eq. (3.7) shows that the original function times a constant must be equal and opposite in sign to the second derivative of the function. Very few common functions meet this requirement, but $\sin bx$, $\cos bx$, and an exponential function, e^{ibx}, do have this property.

$$\frac{d^2}{dx^2}(\sin bx) = -b^2 \sin bx$$

$$\frac{d^2}{dx^2}\left(e^{ibx}\right) = -b^2 e^{ibx}.$$

In the solution shown in Eq. (3.8), A and B are constants and $k = \frac{(2mE)^{1/2}}{\hbar}$.

The appearance of two constants in the solution is normal for a second-order differential equation. These constants must be evaluated using the *boundary conditions*. Boundary conditions are those requirements that must be met because of the physical limits of the system.

In order for the probability of finding the particle to vanish at the walls of the box, we require that ψ be zero at the boundaries. That is, $\psi = 0$ for $x \leq 0$ or $x \geq a$. At $x = 0$,

$$\psi = 0 = A\cos\left[\frac{(2mE)^{1/2}}{\hbar}\cdot 0\right] + B\sin\left[\frac{(2mE)^{1/2}}{\hbar}\cdot 0\right]. \tag{3.9}$$

Now $\sin 0 = 0$ so the last term is 0; $\cos 0 = 1$ so

$$\psi = 0 = A \cdot 1.$$

This can be true only for $A = 0$. Therefore, A must be 0 and the wave function reduces to

$$\psi = B\sin\left[(2mE)^{1/2}x/\hbar\right], \tag{3.10}$$

but we must now evaluate B. Using the requirement that the wave function must vanish at the boundary a, $\psi = 0$ for $x = a$, and

$$\psi = 0 = B\sin\left[(2mE)^{1/2}a/\hbar\right]. \tag{3.11}$$

Now $\sin\theta = 0$ for $\theta = 0°, 180°, 360°, \ldots$, which represents $\theta = n\pi$ rad, where n is an integer. Consequently,

$$(2mE)^{1/2}a/\hbar = n\pi, \tag{3.12}$$

where $n = 1, 2, 3, \ldots$. The value $n = 0$ would lead to $\psi = 0$, which would give a probability of 0 for finding the particle inside the box. Since the particle is required to be somewhere inside the box, the value of $n = 0$ is rejected. We can solve the previous expression to give the allowed energy levels in terms of n,

$$E = \frac{n^2\hbar^2\pi^2}{2ma^2} = \frac{n^2h^2}{8ma^2}, \tag{3.13}$$

where $n = 1, 2, 3, \ldots$, a *quantum number*. Note that the requirement that the wave function must vanish at the boundaries of the box (0 and a) causes the quantization of energy. This occurs because the trigonometric functions vanish only for certain values of θ. Therefore, for a free particle, the energy levels are not quantized, but rather they form a continuum. It is only for the bound (confined or constrained) system that the energy levels are quantized. We should therefore expect that an electron bound in a hydrogen atom should exhibit discrete energy levels. The diverging energy levels for the particle in a one-dimensional box are shown in a graphical way in Figure 3.2. In most systems of a chemical nature (recall the energy level diagram for the hydrogen atom), the energy levels converge. Note also that the energy of the lowest state is not zero and that the lowest state has an energy of $h^2/8ma^2$.

Although we have found the allowed energy levels for the particle and the general form of the wave function, the wave function has not been

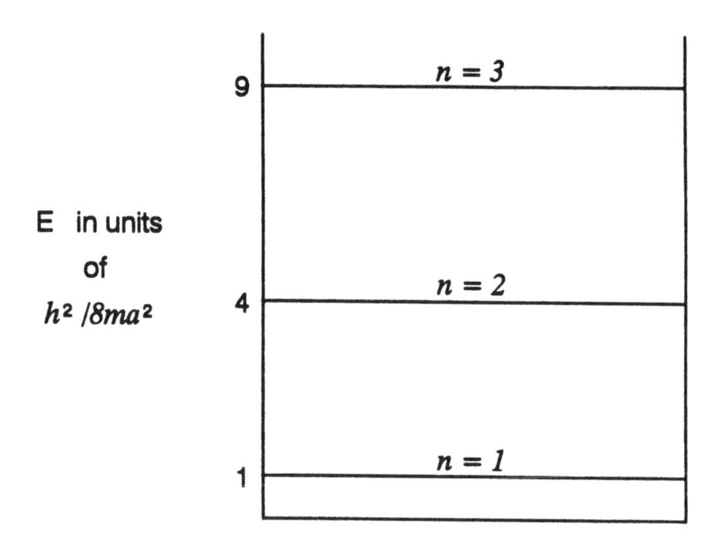

Figure 3.2 ▸ Energy levels for a particle in a one-dimensional box.

normalized. The wave function is normalized when

$$\int_{\text{all space}} \psi^*\psi \, d\tau = 1. \tag{3.14}$$

Therefore, for this problem the integration is over the interval in which x can vary, which is from 0 to a. If we let B be the *normalization constant*,

$$\int_0^a \mathrm{B}^*\psi^*\mathrm{B}\psi \, d\tau = 1 \tag{3.15}$$

and since $\psi = \mathrm{B}\sin\left(\frac{n\pi x}{a}\right)$ and $\mathrm{B}^* = \mathrm{B}$, this can be written as

$$\int_0^a \mathrm{B}^2 \sin^2 (n\pi/a)\, x \, dx = 1. \tag{3.16}$$

This integral (where a is a constant) is a standard form given in tables of integrals as

$$\int \left(\sin^2 ax\right) dx = \frac{1}{2}x - \frac{1}{4a}\sin 2ax, \tag{3.17}$$

and in this problem, after solving for B^2, integration gives

$$\mathrm{B}^2 = \frac{1}{\left.\frac{x}{2}\frac{(\sin 2\pi x/a)}{4n\pi/a}\right|_{x=0}^{x=a}}. \tag{3.18}$$

The denominator evaluates to $a/2$ so that $\mathrm{B} = (2/a)^{1/2}$. The complete normalized wave function can be written as

$$\psi = \left(\frac{2}{a}\right)^{1/2} \sin\left(\frac{n\pi}{a}\right)x. \tag{3.19}$$

Using this wave function, the average or expectation value of position or momentum of the particle can be calculated using the results of Postulate III shown in Chapter 2. Figure 3.3 shows the plots of ψ and ψ^2 for the first few values of n.

We will consider a carbon chain like

$$\text{C=C–C=C–C} \tag{3.20}$$

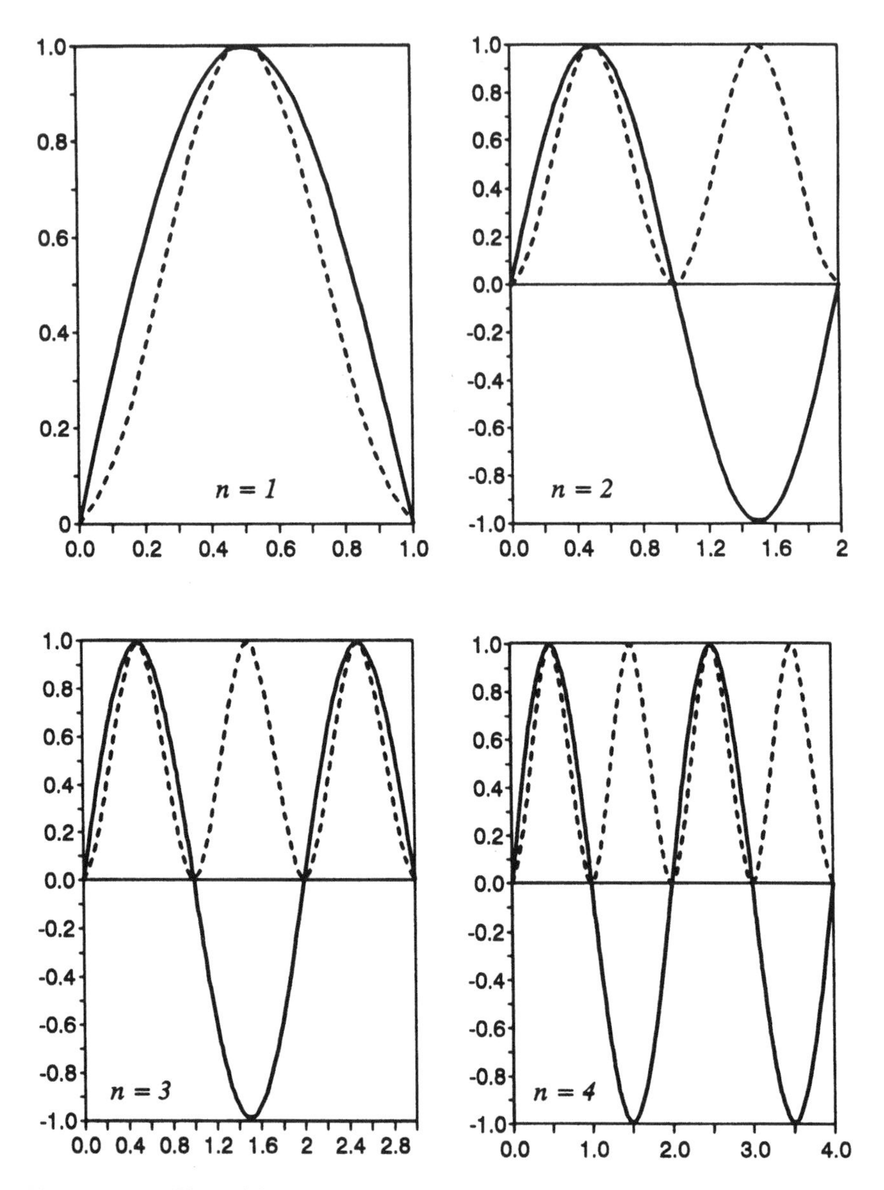

Figure 3.3 ▶ Plots of the wave function (solid line) and the square of the wave function (dashed line) for the first four states of the particle in a box.

as an arrangement where the π electrons can move along the chain. If we take an average bond length of 1.40 Å, the entire chain would be 5.60 Å in length. Therefore, the difference between the $n = 1$ and the $n = 2$ states would be

$$E = \frac{2^2h^2}{8ma^2} - \frac{1^2h^2}{8ma^2} = \frac{3h^2}{8ma^2} \tag{3.21}$$

$$E = \frac{3\left(6.63 \times 10^{-27}\ \text{erg s}\right)^2}{8\left(9.10 \times 10^{-28}\text{g}\right) \times \left(5.60 \times 10^{-8}\text{cm}\right)^2} \tag{3.22}$$

$$= 5.78 \times 10^{-12}\ \text{erg}.$$

This corresponds to a wavelength of 344 nm, and the actual maximum in the absorption spectrum of 1,3-pentadiene is found at 224 nm. Although this is not close agreement, the simple model does predict absorption in the ultraviolet region of the spectrum.

Although this model is of limited usefulness for physical problems, the methodology that it shows is valuable for illustrating the quantum mechanical way of doing things. A few observations are in order here. First, the energy of a confined particle is quantized. Application of the boundary conditions leads to the quantization of energy. Also, energy increases as the square of the quantum number describing the state. The energy also increases as the mass of the particle decreases. This has implications for the confinement of a particle having the mass of an electron to a region the size of an atomic nucleus (approximately 10^{-13} cm). Therefore, the β^- particles emitted during beta decay could not exist in the nucleus prior to the actual decay. For example, confining an electron this way would require a lowest state energy of

$$E = \frac{1^2h^2}{8ma^2} = \frac{1\left(6.63 \times 10^{-27}\ \text{erg s}\right)^2}{8\left(9.10 \times 10^{-28}\ \text{g}\right) \times \left(10^{-13}\ \text{cm}\right)^2} = 0.604\ \text{erg}.$$

Using the conversion factor 1 erg $= 6.242 \times 10^{11}$ eV, this amounts to 3.77×10^{11} eV or 3.77×10^5 MeV!

It is interesting to note that the energy, in addition to being quantized, depends on the quantum number n, and n cannot be zero. Therefore, there is *some* energy for the particle ($E = h^2/8ma^2$ when $n = 1$) even in the lowest state. This is known as the *zero-point energy*. As we shall see later, other systems (the hydrogen atom, the harmonic oscillator, etc.) have a zero-point energy as well.

The second important result is that *one quantum number arises from the solution of an equation for a one-dimensional system.* This quantum number arises as a mathematical restriction or condition rather than as an assumption as it did in the case of the Bohr treatment of the hydrogen atom.

It will not be surprising when it turns out that a two-dimensional system gives rise to two quantum numbers, three dimensions to three quantum numbers, and so on. The particle in the one-dimensional box can serve as a useful first approximation for electrons moving along a conjugated hydrocarbon chain.

3.2 Separation of Variables

Suppose a differential equation can be written as

$$\frac{\partial^2 U}{\partial x^2} - \frac{\partial U}{\partial y} = 0. \tag{3.23}$$

The solution of the equation requires finding the solution that is a function of x and y, $U = U(x, y)$. Let us now *assume* that a solution exists such that $U(x, y) = X(x)Y(y)$, where X and Y are functions of x and y, respectively, so that $U = XY$. Equation (3.23) becomes

$$\frac{\partial (XY)}{\partial y} = \frac{\partial^2 (XY)}{\partial x^2}. \tag{3.24}$$

Now X is *not* a function of y, and Y is not a function of x, so we can treat X and Y as constants to give

$$\frac{X \partial Y}{\partial y} = \frac{Y \partial^2 X}{\partial x^2} \tag{3.25}$$

This equation can be rearranged to give

$$\frac{X''}{X} = \frac{Y'}{Y}. \tag{3.26}$$

Each side of the equation is a constant with respect to the other since one is a function of x and the other is a function of y. Therefore, we can write

$$\frac{X''}{X} = C \text{ and } \frac{Y'}{y} = C. \tag{3.27}$$

Each differential equation can now be solved independently of the other to obtain $X(x)$ and $Y(y)$. The desired solution is $U(x, y) = X(x)Y(y)$. The separation of variables technique is commonly used in solving partial differential equations. Solution of the wave mechanical equation for the hydrogen atom also requires this technique, as we will see in the next chapter.

3.3 The Particle in a Three-Dimensional Box

The particle in a three-dimensional box model illustrates additional aspects of the quantum mechanical methods. In this problem, we have a particle in a box of dimensions a, b, and c in the x, y, and z directions, respectively, as shown in Figure 3.4. As before, we will take the potential energy inside the box to be zero, but outside the box we will set $V = \infty$. Therefore,

$$V_{\text{total}} = V_x + V_y + V_z \tag{3.28}$$

and

$$V = 0 \quad \text{for} \quad \begin{array}{c} 0 < x < a \\ 0 < y < b \\ 0 < z < c \end{array} \quad \text{and} \quad V = \infty \quad \text{for} \quad \begin{array}{c} 0 > x > a \\ 0 > y > b \\ 0 > z > c. \end{array}$$

Inside the box, the Hamiltonian can now be written as

$$H = T + V = \frac{p^2}{2m} + 0. \tag{3.29}$$

The Hamiltonian operator can be written using the kinetic energy expressed in terms of the momentum as in Chapter 2.

$$-\frac{\hbar^2}{2m}\left(\frac{\partial^2}{\partial x^2} + \frac{\partial^2}{\partial y^2} + \frac{\partial^2}{\partial z^2}\right) = -\frac{\hbar^2}{2m}\nabla^2. \tag{3.30}$$

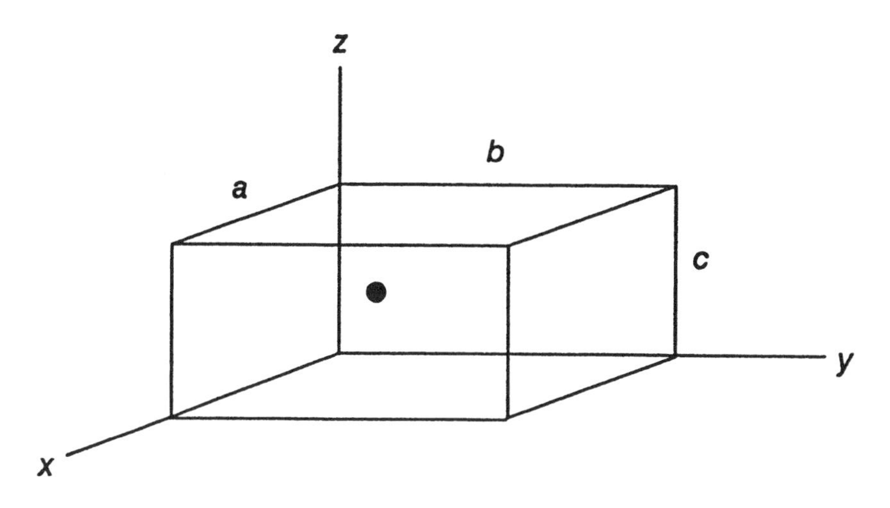

Figure 3.4 ▶ A particle in a three-dimensional box.

Therefore, the equation $\hat{H}\psi = E\psi$ becomes

$$-\frac{\hbar^2}{2m}\nabla^2\psi = E\psi \tag{3.31}$$

or

$$\nabla^2\psi + \frac{2m}{\hbar^2}E\psi = 0. \tag{3.32}$$

This is a partial differential equation in three variables (x, y, and z). The simplest method to solve such an equation and the one that should be tried first is the method of *separation of variables*. We will see that this is also the case in the hydrogen atom problem in the next chapter. To separate the variables, it is assumed that the desired solution, ψ, can be factored into three functions, each of which is a function of one variable only. In other words, we assume that

$$\psi(x, y, z) = X(x)\,Y(y)\,Z(z)\,. \tag{3.33}$$

This product of three functions is now written in place of ψ, but we will use the simplified notation $X = X(x)$, etc.:

$$\frac{\partial^2 XYZ}{\partial x^2} + \frac{\partial^2 XYZ}{\partial y^2} + \frac{\partial^2 XYZ}{\partial z^2} + \frac{2m}{\hbar^2}E\,(XYZ) = 0. \tag{3.34}$$

Since YZ is not a function of x, XZ is not a function of y, and XY is not a function of z, we can remove them from the derivatives to give

$$YZ\frac{\partial^2 X}{\partial x^2} + XZ\frac{\partial^2 Y}{\partial y^2} + XY\frac{\partial^2 Z}{\partial z^2} + \frac{2m}{\hbar^2}E\,(XYZ) = 0. \tag{3.35}$$

If Eq. (3.35) is divided by XYZ, we obtain

$$\frac{1}{X}\frac{\partial^2 X}{\partial x^2} + \frac{1}{Y}\frac{\partial^2 Y}{\partial y^2} + \frac{1}{Z}\frac{\partial^2 Z}{\partial z^2} = -\frac{2m}{\hbar^2}E. \tag{3.36}$$

Since each term on the left-hand side of Eq. (3.36) is a function of only one variable, each will be independent of any change in the other two variables. Each term must be equal to a constant, which we will call $-k^2$. Since there must be three such constants, we will call the constants $-k_x^2$, $-k_y^2$, and $-k_z^2$ for the x, y, and z directions, respectively. We thus have three equations that can be written using ordinary derivatives as

$$\frac{1}{X}\frac{d^2X}{dx^2} = -k_x^2 \tag{3.37}$$

$$\frac{1}{Y}\frac{d^2Y}{dy^2} = -k_y^2 \tag{3.38}$$

$$\frac{1}{Z}\frac{d^2Z}{dz^2} = -k_z^2. \tag{3.39}$$

The sum of the three constants must be equal to the right-hand side of Eq. (3.36), which is $-2mE/\hbar^2$. Therefore,

$$k_x^2 + k_y^2 + k_z^2 = \frac{2mE}{\hbar^2}. \tag{3.40}$$

The energy is the sum of the contributions from each degree of freedom in the x, y, and z coordinates:

$$E = E_x + E_y + E_z. \tag{3.41}$$

Therefore,

$$k_x^2 + k_y^2 + k_z^2 = \frac{2m}{\hbar^2}(E_x + E_y + E_z). \tag{3.42}$$

Since the energy associated with the degree of freedom in the x direction is not dependent on the y and z coordinates, we can separate Eq. (3.42) to give

$$k_x^2 = \frac{2mE_x}{\hbar^2}; \quad k_y^2 = \frac{2mE_y}{\hbar^2}; \quad k_z^2 = \frac{2mE_z}{\hbar^2}. \tag{3.43}$$

The first of the equations, Eq. (3.37), can be written as

$$\frac{d^2X(x)}{dx^2} + k_x^2X(x) = 0. \tag{3.44}$$

This equation is of the same form as Eq. (3.7) so the solution can be written directly as

$$X(x) = \sqrt{\frac{2}{a}}\sin\frac{n_x\pi}{a}x. \tag{3.45}$$

Similarly, the other two equations yield

$$Y(y) = \sqrt{\frac{2}{b}}\sin\frac{n_y\pi}{b}y, \tag{3.46}$$

$$Z(z) = \sqrt{\frac{2}{c}}\sin\frac{n_z\pi}{c}z. \tag{3.47}$$

The general solution can be written as the product of the three partial solutions (the assumption made earlier)

$$\psi(x, y, z) = X(x)\,Y(y)\,Z(z) \tag{3.48}$$

$$\psi(x, y, z) = \sqrt{\frac{2}{a}} \sin \frac{n_x \pi}{a} x \cdot \sqrt{\frac{2}{b}} \sin \frac{n_y \pi}{b} y \cdot \sqrt{\frac{2}{c}} \sin \frac{n_z \pi}{c} z, \tag{3.49}$$

where n_x, n_y, and n_z, are the quantum numbers for the x, y, and z components of energy, respectively. The general solution can be simplified somewhat to give

$$\psi(x, y, z) = \sqrt{\frac{8}{abc}} \sin \frac{n_x \pi}{a} x \sin \frac{n_y \pi}{b} y \sin \frac{n_z \pi}{c} z. \tag{3.50}$$

We can draw some analogies to the particle in the one-dimensional box. First, we will find expressions for the energies using Eq. (3.43).

$$\frac{n_x \pi}{a} = k_x = \sqrt{\frac{2mE_x}{\hbar^2}}. \tag{3.51}$$

We can now write the expression for the energy levels based on the x component as

$$E_x = \frac{\hbar^2 n_x^2 \pi^2}{2ma^2} = \frac{h^2 n_x^2}{8ma^2}. \tag{3.52}$$

The equivalent expressions for the energy based on the y and z directions are

$$E_y = \frac{\hbar^2 n_y^2 \pi^2}{2mb^2} \quad \text{and} \quad E_z = \frac{\hbar^2 n_z^2 \pi^2}{2mc^2} \tag{3.53}$$

The total energy, E, is

$$E = E_x + E_y + E_z = \frac{\hbar^2 \pi^2}{2m} \left(\frac{n_x^2}{a^2} + \frac{n_y^2}{b^2} + \frac{n_z^2}{c^2} \right), \tag{3.54}$$

where $n_x = 1, 2, 3, \ldots$; $n_y = 1, 2, 3, \ldots$; and $n_z = 1, 2, 3 \ldots$.

It should be noted that one quantum number has been introduced for each degree of freedom of the system, the three coordinates of the particle.

Therefore, the energy is dependent on n_x, n_y, and n_z. If we assume that the box is cubic, $a = b = c$. Therefore, the denominators of the fractions are identical and the lowest energy will occur when the numerators have the smallest values (which occurs when all of the quantum numbers are 1). This state can be designated as the 111 state, where the digits indicate the values of the quantum numbers n_x, n_y, and n_z, respectively. The state of next lowest energy would be with two of the quantum numbers being 1 and the other being 2. One way that this could occur would be with $n_x = 2$ and $n_y = n_z = 1$. In this case, the energy would be equal to $6\hbar^2\pi^2/2ma^2$ or $6h^2/8ma^2$ when $\hbar = h/2\pi$ is substituted. However, this state, designated as the 211 state, has the same energy as the 121 and 112 states. Therefore, these states are *degenerate* in the case where $a = b = c$. However, if the dimensions of the box are not equal, then the 211, 121, and 112 states are not degenerate. It is easy to see that if one of the dimensions is twice another (or in some other appropriate relationship), the energies might still happen to be degenerate simply because of the relationship of a, b, and c as the quantum numbers are assigned different values. Such a situation is known as *accidental degeneracy*.

The energy level diagram that results when $a = b = c$ is shown in Figure 3.5. The states are indicated in terms of the quantum numbers n_x, n_y, and n_z (e.g., 112, 123, and 322), and the degeneracy is given after the combination of quantum numbers.

When $a \neq b \neq c$, the energies must be calculated by using the actual dimensions of the box. Choosing unequal values for a, b, and c and then using several integers for n_x, n_y, and n_z will quickly enable one to see the nondegeneracy.

When potassium vapor is passed over a crystal of KCl, the crystal takes on a color. It can be shown that as a result of the reaction,

$$K(\text{vap}) \rightarrow K + (\text{crystal}) + e^-(\text{anion site}), \tag{3.55}$$

the electrons occupy anion sites in the KCl lattice. In reality, the electrons are distributed over the cations that surround the lattice site. The centers where the electrons reside are responsible for the absorption of light, which results in the crystal being colored. Such centers are called *f-centers* because of the German word *farbe*, which means "color." When other alkali metals are added in the same way to the corresponding alkali halides, color centers are also produced. As a very crude approximation, the electrons in anion sites can be treated as particles in three-dimensional boxes. It is interesting to note that the wavelength of the light absorbed

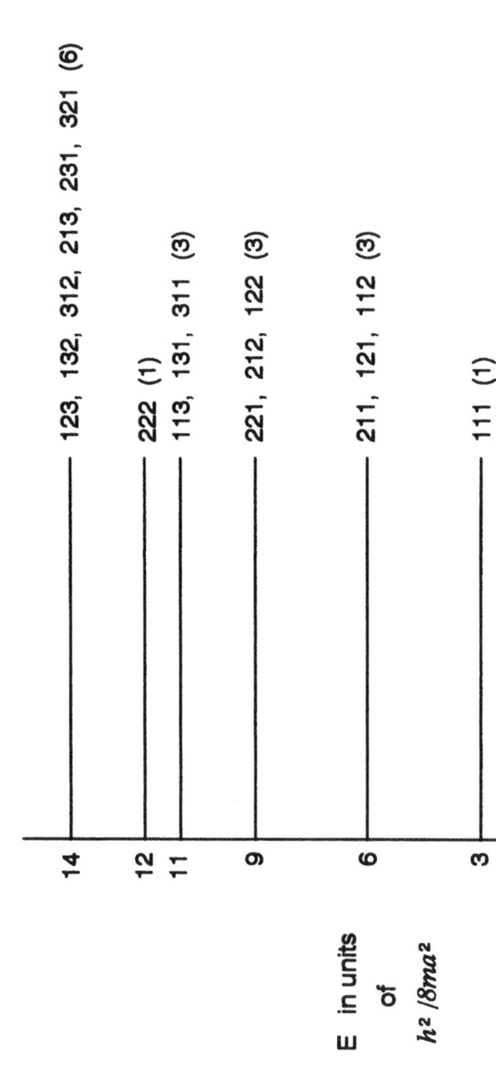

Figure 3.5 ▶ Energy levels for a particle in a three-dimensional cubic box.

depends on the nature of the crystal lattice, and the maxima in the absorptions for several crystals are as follow:

Crystal	Absorption maximum (erg)
LiCl	4.96×10^{-12}
NaCl	4.32×10^{-12}
KCl	3.52×10^{-12}
RbCl	3.20×10^{-12}
LiBr	4.32×10^{-12}
NaBr	3.68×10^{-12}

For a given chloride compound, the size of the anion site (where the electron resides) is dependent on the size of the cation. Since the energies for a particle in a three-dimensional box are inversely related to the size of the box, we expect the greatest difference between energy levels for LiCl. Accordingly, the absorption energy is highest for LiCl for which the anion site is smallest. In fact, the series from LiCl to RbCl shows this trend clearly based on the size of the anion site. Therefore, this phenomenon shows a correlation that would be predicted when the particle in a three-dimensional box model is employed.

References for Further Reading

- Alberty, R. A., and Silbey, R. J. (1996). *Physical Chemistry*, 2nd ed. Wiley, New York. The chapters on quantum mechanics are the equivalent of some books on the subject. A thorough introduction to and treatment of the basic models.
- Eyring, H., Walter, J., and Kimball, G. E. (1944). *Quantum Chemistry*. Wiley, New York. One of the two true (high-level) classics in the applications of quantum mechanics to chemistry.
- Laidler, K. J., and Meiser, J. H. (1982). *Physical Chemistry*, Chap. 11. Benjamin–Cummings, Menlo Park, CA. A clear introduction to the basic methods of quantum mechanics.
- Pauling, L., and Wilson, E. B. (1935). *Introduction to Quantum Mechanics*. McGraw–Hill, New York. The other classic in quantum mechanics aimed primarily at chemistry. Still a definitive work. Now available in an inexpensive, widely available reprint from Dover.
- Sherwin, C. W. (1959). *Introduction to Quantum Mechanics*. Holt, Rinehart, & Winston, New York. One of the best overall accounts of quantum mechanics and a standard text.

Problems

1. Solve the equation $y'' + ay = 0$ subject to the boundary conditions $y(0) = y(\pi) = 0$.

2. If a hexatriene molecule absorbs light of 2500 Å to change a π electron from $n = 1$ to $n = 2$, what is the length of the molecule?

3. What would be the translational energies of the first two levels for a hydrogen molecule confined to a length of 10 cm?

4. What would be the length of a one-dimensional box necessary for the separation between the first two energy levels for a proton to be 2.00 eV?

5. Calculate the probability of finding the particle in a one-dimensional box of length a in the interval $0.100a$ to $0.250a$.

6. Planck's constant is the fundamental quantum of action (energy $\times$ time). Explain how, as the fundamental quantum of action approaches zero, the behavior of a particle in a box becomes classical.

7. Calculate the average value of the x coordinate of a particle in a one-dimensional box.

8. Consider an atomic nucleus to be a potential box 10^{-13} cm in diameter. If a neutron falls from $n = 2$ to $n = 1$, what energy is released? If this energy is emitted as a photon, what wavelength will it have? In what region of the electromagnetic spectrum will it be observed?

9. Consider a particle of mass m moving in a planar circular path of length l. Assume that the potential for the particle along the path is zero while the potential for the particle to not be on the path is infinite to confine the particle to the path.
   - **(a)** Set up the wave equation for this model.
   - **(b)** Solve the wave equation to get a general form of the solution.
   - **(c)** Use the fact that the solution for any points x and $(l + x)$ must be equal to simplify the solution.

10. When sodium dissolves in liquid ammonia, some dissociation occurs:

$$\text{Na} \longrightarrow \text{Na}^+(\text{solvated}) + e^-(\text{solvated}).$$

The solvated electron can be treated as a particle in a three-dimensional box. Assume that the box is cubic with an edge length of 1.55×10^{-7} cm and suppose that excitation occurs in all directions simultaneously for the lowest state to the first excited state. What wavelength of radiation would the electron absorb? Would the solution be colored?

11. What size would a one-dimensional box holding an electron have to be in order for it to have the same energy as a hydrogen molecule would have in a box of length 10 Å?

12. Suppose a helium atom is in a box of length 50 Å. Calculate the energy at few distances and sketch the energy as a function of box length.

13. Suppose a particle in a three-dimensional box has an energy of $\frac{14h^2}{8ma^2}$. If $a = b = c$ for the box, what is the degeneracy of this state?

14. Show that the wave functions for the first two energy levels of a particle in a one-dimensional box are orthogonal.

15. Consider an electron in the π bond in ethylene as a particle in a one-dimensional box of length 133 pm. What is the energy difference between the first two energy levels? In what region of the electromagnetic spectrum would a photon emitted as the electron falls from the first excited to the ground state be observed?

16. An electron trapped in a three-dimensional lattice defect (vacant anion site) of a crystal can be considered as a particle in a three-dimensional box. If the length of the box in each dimension is 200 nm, what would be the difference between the first two allowed states? What would be the effect on the energies of the first two allowed states if the defect site were 200 nm in length in the x and y directions but 250 nm in the z direction?

▶ Chapter 4

The Hydrogen Atom

The work of de Broglie showed that a moving particle has a wave character. In Chapter 2 we saw that this can lead to an adaptation of an equation that is known to apply to vibrations in three dimensions to give an equation that describes the electron in a hydrogen atom as a three-dimensional wave. Although this was illustrated in the last chapter, we did not solve the resulting equation. We now address that problem for which Erwin Schrödinger received the Nobel prize. We suspect, therefore, that the solution is not a trivial problem!

4.1 Schrödinger's Solution to the Hydrogen Atom Problem

A hydrogen atom can be described in terms of polar coordinates as shown in Figure 4.1. As the electron circles around the nucleus, the system (the proton and the electron) rotates around the center of gravity. For the rotating system, we can write the reduced mass, μ, as

$$\frac{1}{\mu} = \frac{1}{m_{\mathrm{e}}} + \frac{1}{m_{\mathrm{p}}} \tag{4.1}$$

or, solving for μ,

$$\mu = \frac{m_{\mathrm{e}} m_{\mathrm{p}}}{m_{\mathrm{e}} + m_{\mathrm{p}}}. \tag{4.2}$$

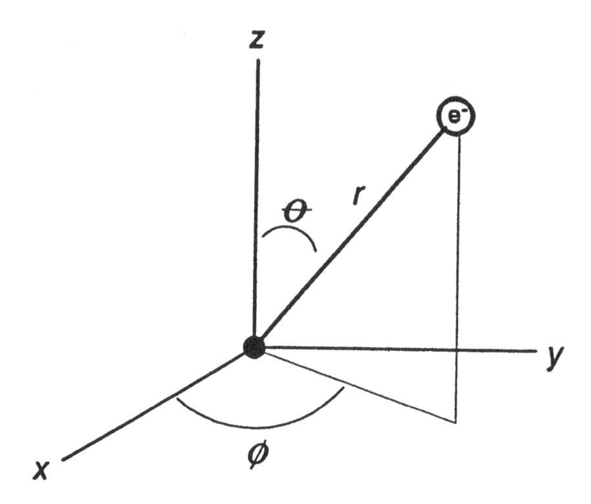

Figure 4.1 ▶ Coordinate system for the hydrogen atom in polar coordinates.

Since the mass of the electron is so much less than that of the proton, $m_e + m_p \approx m_p$ and $\mu = m_e$. Assuming that the nucleus is stationary and that the electron does all of the moving (known as the Born-Oppenheimer approximation) leads to the same result. Therefore, we will assume that this approximation can be used although μ is indicated.

As we saw in the last chapter, the Hamiltonian can be written as the sum of the potential and kinetic energies,

$$H = T + V, \tag{4.3}$$

and $T = p^2/2\mu$. The potential energy, V, for the interaction of the electron with the proton is $-e^2/r$. Therefore, the Hamiltonian, H, for the hydrogen atom is

$$H = \frac{p^2}{2\mu} - \frac{e^2}{r}. \tag{4.4}$$

In operator form, $-e^2/r$ is unchanged because e is a *constant* and r is a *coordinate*. We have already seen that the kinetic energy in operator form (when $\hbar = h/2\pi$, usually called "h-bar") is written as

$$\hat{T} = -\frac{\hbar^2}{2\mu}\nabla^2. \tag{4.5}$$

Therefore, the Hamiltonian operator, $\hat{H}$, is

$$\hat{H} = -\frac{\hbar^2}{2\mu}\nabla^2 - \frac{e^2}{r}$$

and

$$\hat{H}\psi = E\psi. \tag{4.6}$$

Substituting for the Hamiltonian operator, $\hat{H}$, we obtain

$$-\frac{\hbar^2}{2\mu}\nabla^2\psi - \frac{e^2}{r}\psi = E\psi \tag{4.7}$$

or

$$\nabla^2\psi + \frac{2\mu}{\hbar^2}(E - V)\psi = 0. \tag{4.8}$$

When the Laplacian operator, ∇^2, is written out, we obtain

$$\left(\frac{\partial^2}{\partial x^2} + \frac{\partial^2}{\partial y^2} + \frac{\partial^2}{\partial z^2}\right)\psi + \frac{2\mu}{\hbar^2}(E - V)\psi = 0. \tag{4.9}$$

This is a partial differential equation in three variables and we suspect that the separation of variables technique must be applied to solve it. However,

$$r = (x^2 + y^2 + z^2)^{1/2},$$

and there is no way to separate the variables. The trick now is to change coordinate systems to get variables (coordinates) that can be separated.

The transformation that is made is to describe the hydrogen atom in terms of *polar* coordinates. Figure 4.1 shows the coordinate system and the relationship between the Cartesian and the polar coordinates. The knotty part of this problem now consists of transforming

$$\frac{\partial^2}{\partial x^2} + \frac{\partial^2}{\partial y^2} + \frac{\partial^2}{\partial z^2} \text{ into } f(r, \theta, \phi). \tag{4.10}$$

That transformation is laborious, but the result is

$$\nabla^2 = \frac{1}{r^2}\frac{\partial}{\partial r}\left(r^2\frac{\partial}{\partial r}\right) + \frac{1}{r^2 \sin\theta}\frac{\partial}{\partial\theta}\left(\sin\theta\frac{\partial}{\partial\theta}\right) + \frac{1}{r^2\sin^2\theta}\frac{\partial^2}{\partial\phi^2}. \tag{4.11}$$

Now, the Schrödinger equation becomes

$$\frac{1}{r^2}\frac{\partial}{\partial r}\left(r^2\frac{\partial}{\partial r}\psi\right) + \frac{1}{r^2 \sin\theta}\frac{\partial}{\partial\theta}\left(\sin\theta\frac{\partial}{\partial\theta}\psi\right) + \frac{1}{r^2\sin^2\theta}\frac{\partial^2}{\partial\phi^2}\psi + \frac{2\mu}{\hbar^2}\left(\frac{e^2}{r} + E\right)\psi = 0, \tag{4.12}$$

which contains only the variables r, θ, and ϕ.

In this case, we assume that $\psi(r, \theta, \phi) = R(r)\Theta(\theta)\Phi(\phi)$ and make that substitution. For simplicity, we will write the partial solutions as R, Θ, and Φ without showing the functionality, $R(r)$, etc. The Schrödinger equation becomes

$$\frac{1}{r^2}\frac{\partial}{\partial r}r^2\frac{\partial R\Theta\Phi}{\partial r} + \frac{1}{r^2 \sin\theta}\frac{\partial}{\partial\theta}\left(\sin\theta\frac{\partial R\Theta\Phi}{\partial\theta}\right) + \frac{1}{r^2\sin^2\theta}\frac{\partial^2 R\Theta\Phi}{\partial\phi^2} + \frac{2\mu}{\hbar^2}\left(\frac{e^2}{r} + E\right)R\Theta\Phi = 0. \tag{4.13}$$

However, since Θ and Φ are not functions of r, they can be removed (as constants) from the differentiation. The same action is possible in other terms of the equation involving the other variables. The wave equation then becomes

$$\frac{\Theta\Phi}{r^2}\frac{\partial}{\partial r}r^2\frac{\partial R}{\partial r} + \frac{R\Phi}{r^2\sin\theta}\frac{\partial}{\partial\theta}\left(\sin\theta\frac{\partial\Theta}{\partial\theta}\right) + \frac{R\Theta}{r^2\sin^2\theta}\frac{\partial^2\Phi}{\partial\phi^2} + \frac{2\mu}{\hbar^2}\left(\frac{e^2}{r} + E\right)R\Theta\Phi = 0. \tag{4.14}$$

We now divide both sides of Eq. (4.14) by $R\Theta\Phi$ and multiply by $r^2\sin^2\theta$:

$$\frac{\sin^2\theta}{R}\frac{\partial}{\partial r}\left(r^2\frac{\partial R}{\partial r}\right) + \frac{\sin\theta}{\Theta}\frac{\partial}{\partial\theta}\left(\sin\theta\frac{\partial\Theta}{\partial\theta}\right) + \frac{1}{\Phi}\frac{\partial^2\Phi}{\partial\phi^2} + \frac{2\mu r^2\sin^2\theta}{\hbar^2}\left(\frac{e^2}{r} + E\right) = 0. \tag{4.15}$$

Inspection shows that of the four terms on the left-hand side of the equation, there is no functional dependence on ϕ except in the third term. Therefore, with respect to the other variables, the third term can be treated as a constant. For convenience, we will set it equal to $-m^2$. This m is *not* the same as the electron mass used in $2m/\hbar^2$. Therefore, we obtain

$$\frac{1}{\Phi}\frac{\partial^2\Phi}{\partial\phi^2} = -m^2 \tag{4.16}$$

or

$$\frac{\partial^2\Phi}{\partial\phi^2} + m^2\Phi = 0. \tag{4.17}$$

This represents the "ϕ equation" and we have partially separated the variables.

We can now write the wave equation as

$$\frac{\sin^2\theta}{R}\frac{\partial}{\partial r}\left(r^2\frac{\partial R}{\partial r}\right)+\frac{\sin\theta}{\Theta}\frac{\partial}{\partial\theta}\left(\sin\theta\frac{\partial\Theta}{\partial\theta}\right)-m^2 \tag{4.18}$$
$$+\frac{2\mu r^2\sin^2\theta}{\hbar^2}\left(\frac{e^2}{r}+E\right)=0.$$

If we divide the equation by $\sin^2\theta$ and rearrange, we obtain

$$\frac{1}{R}\frac{\partial}{\partial r}\left(r^2\frac{\partial R}{\partial r}\right)+\frac{2\mu r^2}{\hbar^2}\left(\frac{e^2}{r}+E\right)+\frac{1}{\Theta\sin\theta}\frac{\partial}{\partial\theta} \tag{4.19}$$
$$\left(\sin\theta\frac{\partial\Theta}{\partial\theta}\right)-\frac{m^2}{\sin^2\theta}=0.$$

Inspection of this equation shows that the first two terms contain the functional dependence on r and the last two terms reflect the dependence on θ. As before, we will set the two terms equal to a constant, β, so that

$$\frac{1}{R}\frac{\partial}{\partial r}\left(r^2\frac{\partial R}{\partial r}\right)+\frac{2\mu r^2}{\hbar^2}\left(\frac{e^2}{r}+E\right)=-\beta \tag{4.20}$$

$$\frac{1}{\Theta\sin\theta}\frac{\partial}{\partial\theta}\left(\sin\theta\frac{\partial\Theta}{\partial\theta}\right)-\frac{m^2}{\sin^2\theta}=-\beta. \tag{4.21}$$

If we multiply Eq. (4.20) by R and Eq. (4.21) by Θ, we obtain

$$\frac{\partial}{\partial r}\left(r^2\frac{\partial R}{\partial r}\right)+\frac{2\mu r^2}{\hbar^2}\left(\frac{e^2}{r}+E\right)R+R\beta=0 \tag{4.22}$$

$$\frac{1}{\sin\theta}\frac{\partial}{\partial\theta}\left(\sin\theta\frac{\partial\Theta}{\partial\theta}\right)-\frac{m^2}{\sin^2\theta}\Theta+\beta\Theta=0. \tag{4.23}$$

The variables r, θ, and ϕ have now been separated and the second-order partial differential equation in three variables has been transformed into three second-order differential equations, each in one variable.

Solving the three equations is now the task. Only the equation involving ϕ is simple in its solution because it is of the same form as the equation for the particle in a box problem. The solution of the "ϕ equation" can be written as

$$\Phi\left(\phi\right)=\frac{1}{\sqrt{2\pi}}\mathrm{e}^{im\phi}. \tag{4.24}$$

The solution of the overall equation can be written as

$$\psi\left(r,\theta,\phi\right)=\frac{1}{\sqrt{2\pi}}\mathrm{e}^{im\phi}R\left(r\right)\Theta\left(\theta\right). \tag{4.25}$$

We will now turn our attention to the equation involving θ:

$$\frac{1}{\sin\theta}\frac{\partial}{\partial\theta}\left(\sin\theta\frac{\partial\Theta}{\partial\theta}\right)-\frac{m^2\Theta}{\sin^2\theta}+\beta\Theta=0. \tag{4.26}$$

This equation can be put in the form

$$\frac{d}{\sin\theta\, d\theta}\left(\frac{\sin^2\theta}{\sin\theta\, d\theta}d\Theta\right)-\frac{m^2\Theta}{\sin^2\theta}+\beta\Theta=0. \tag{4.27}$$

The standard method for solving this equation is to make the transformations

$$\begin{aligned} u &= \cos\theta \quad \text{so that} \quad du = -\sin\theta\, d\theta \\ \cos^2\theta &= u^2 = 1-\sin^2\theta \\ \sin^2\theta &= 1-u^2. \end{aligned}$$

Substituting for $\sin^2\theta$ and $\sin\theta\, d\theta$ in Eq. (4.27), we obtain

$$\frac{d}{du}\left(\frac{(1-u^2)}{du}d\Theta\right)-\frac{m^2\Theta}{1-u^2}+\beta\Theta=0, \tag{4.28}$$

which can be written as

$$\left(1-u^2\right)\frac{d^2\Theta}{du^2}-2u\frac{d\Theta}{du}+\left(\beta-\frac{m^2}{1-u^2}\right)\Theta=0. \tag{4.29}$$

This equation is similar in form to a well-known differential equation encountered in advanced mathematics. That equation,

$$\begin{aligned} &\left(1-z^2\right)\frac{d^2P_l^{|m|}\left(z\right)}{dz^2}-2z\frac{dP_l^{|m|}\left(z\right)}{dz}+l\left(l+1\right) \\ &-\frac{m^2}{1-z^2}P_l^{|m|}\left(z\right)=0, \end{aligned} \tag{4.30}$$

is known as Legendre's equation, and β is equivalent to $l\left(l+1\right)$. Solving equations of this type requires the use of series but rather than getting too involved with mathematics at this point we will delay discussing that technique (see Chapter 6). The series solutions of Legendre's equation are known as the *associated Legendre polynomials* and are written as

$$P_l^{|m|}(\cos\theta), \text{ where } l = 0, 1, 2, \ldots, \text{ and } m = 0, \pm 1, \pm 2, \ldots, \pm l.$$

The first few associated Legendre polynomials are

$$\begin{aligned} &l = 0, m = 0: && \Theta(\theta) = 1\sqrt{2} = \Theta_{0,0} \\ &l = 1, m = 0: && \Theta(\theta) = \sqrt{3/2}\cos\theta = \Theta_{1,0} \\ &l = 1, m = \pm 1: && \Theta(\theta) = \sqrt{3/4}\sin\theta = \Theta_{1,\pm 1} \\ &l = 2, m = 0: && \Theta(\theta) = \sqrt{5/8}\left(3\cos^2\theta - 1\right) = \Theta_{2,0}. \end{aligned}$$

The equation that is a function of r is known as the *radial* equation and can be put in the form

$$\frac{1}{r^2}\frac{d}{dr}r^2\frac{dR}{dr} + \frac{2\mu}{\hbar^2}\left(\frac{e^2}{r} + E\right)R - \frac{l(l+1)}{r^2}R = 0. \tag{4.31}$$

This equation can be put in the general form

$$xu'' + u'(2l+2) + (-l-1+n)u = 0, \tag{4.32}$$

which can be solved only when $n \geq l + 1$. This equation is known as *Laguerre's equation* and the solutions are the Laguerre polynomials. We saw earlier that $l = 0, 1, 2, \ldots$, so it is apparent that $n = 1, 2, 3, \ldots$. For example, if $n = 3$, l can take on the values 0, 1, and 2. This gives the familiar restrictions on the quantum numbers that you should have learned in general chemistry (see also Chapter 5):

n = principal quantum number $= 1, 2, 3, \ldots$
l = orbital angular momentum quantum number $= 0, 1, 2, \ldots, (n-1)$
m = magnetic quantum number $= 0, \pm 1, \pm 2, \ldots, \pm l$.

The spin quantum number, s (which is equal to $\pm(1/2)\hbar$), is a property of the electron since it has an intrinsic spin.

The solutions of the equations involving ϕ and θ are combined by multiplication to give the complete angular dependence of the wave functions. These angular functions are known as the *spherical harmonics*, $Y_{l,m}(\theta, \phi)$. Solutions of the equation involving r are called the *radial wave functions*, $R_{n,l}(r)$, and the overall solutions are $R_{n,l}(r)Y_{l,m}(\theta, \phi)$. Table 4.1 gives the

TABLE 4.1 ▶ Complete Wave Functions for the Hydrogen-like Species[a]

$$\psi_{1s} = \frac{1}{\pi^{1/2}}\left(\frac{Z}{a}\right)^{3/2} e^{-Zr/a}$$

$$\psi_{2s} = \frac{1}{4(2\pi)^{1/2}}\left(\frac{Z}{a}\right)^{3/2}\left(2 - \frac{Zr}{a}\right) e^{-Zr/2a}$$

$$\psi_{2p_z} = \frac{1}{4(2\pi)^{1/2}}\left(\frac{Z}{a}\right)^{5/2} r\, e^{-Zr/2a} \cos\theta$$

$$\psi_{2p_x} = \frac{1}{4(2\pi)^{1/2}}\left(\frac{Z}{a}\right)^{5/2} r\, e^{-Zr/2a} \sin\theta \cos\varphi$$

$$\psi_{2p_y} = \frac{1}{4(2\pi)^{1/2}}\left(\frac{Z}{a}\right)^{5/2} r\, e^{-Zr/2a} \sin\theta \sin\varphi$$

$$\psi_{3s} = \frac{1}{81(3\pi)^{1/2}}\left(\frac{Z}{a}\right)^{3/2}\left(27 - 18\frac{Zr}{a} + 2\frac{Z^2r^2}{a^2}\right) e^{-Zr/3a}$$

$$\psi_{3p_z} = \frac{2^{1/2}}{81\pi^{1/2}}\left(\frac{Z}{a}\right)^{5/2}\left(6 - \frac{Zr}{a}\right) r\, e^{-Zr/3a} \cos\theta$$

$$\psi_{3p_x} = \frac{2^{1/2}}{81\pi^{1/2}}\left(\frac{Z}{a}\right)^{5/2}\left(6 - \frac{Zr}{a}\right) r\, e^{-Zr/3a} \sin\theta \cos\varphi$$

$$\psi_{3p_y} = \frac{2^{1/2}}{81\pi^{1/2}}\left(\frac{Z}{a}\right)^{5/2}\left(6 - \frac{Zr}{a}\right) r\, e^{-Zr/3a} \sin\theta \sin\varphi$$

$$\psi_{3d_{xy}} = \frac{1}{81(2\pi)^{1/2}}\left(\frac{Z}{a}\right)^{7/2} r^2 e^{-Zr/3a} \sin^2\theta \sin 2\varphi$$

$$\psi_{3d_{xz}} = \frac{2^{1/2}}{81\pi^{1/2}}\left(\frac{Z}{a}\right)^{7/2} r^2 e^{-Zr/3a} \sin\theta \cos\theta \cos\varphi$$

$$\psi_{3d_{yz}} = \frac{2^{1/2}}{81\pi^{1/2}}\left(\frac{Z}{a}\right)^{7/2} r^2 e^{-Zr/3a} \sin\theta \cos\theta \sin\varphi$$

$$\psi_{3d_{x^2-y^2}} = \frac{1}{81(2\pi)^{1/2}}\left(\frac{Z}{a}\right)^{7/2} r^2 e^{-Zr/3a} \sin\theta \cos 2\varphi$$

$$\psi_{3d_{z^2}} = \frac{1}{81(6\pi)^{1/2}}\left(\frac{Z}{a}\right)^{7/2} r^2 e^{-Zr/3a}(3\cos^2\theta - 1)$$

[a]The nuclear charge is given by Z, and a is the first Bohr radius, 0.529 Å.

wave functions for the hydrogen-like species. If $Z = 1$, the hydrogen wave functions are indicated.

4.2 Interpreting the Solutions

In Chapter 2, the idea that ψ^2 is related to the probability of finding a particle described by the wave function, ψ, was discussed briefly. In classical physics, the square of the amplitude gives the total energy of a vibrating system (e.g., a vibrating object on a spring or a vibrating string). Similarly, the square of the wave function for an electron is proportional to the amplitude function squared because, in reality, we have represented an electron as a de Broglie wave by means of the Schrödinger equation.

Solving a differential equation to obtain ψ does not uniquely determine a probability because solving such an equation leads to arbitrary constants. In the case of the electron, the particle must be *somewhere*, so we write the probability integral as

$$\int \psi^2 \, d\tau = 1. \tag{4.33}$$

If this relationship does not hold, we adjust the constants included in the wave function using the constant N,

$$\int N\psi \cdot N\psi \, d\tau = 1, \tag{4.34}$$

where N is called a normalization constant. Regardless of whether ψ is positive or negative, the value of ψ^2 is positive so the probability ranges from 0 to 1.

Another interpretation of the square of the wave function is not only possible, it also provides a very useful concept for describing certain properties. The concept being described is that of the density of the electron cloud. If flash photography could capture an electron on film and if we repeated the photography an enormous number of times, we could plot the position of the electron at each instant at which it appeared in a photograph. The result would appear as shown in Figure 4.2.

The area where the dots represent the highest density represents the regions where the electron is found most of the time. If we picture the dots as representing particles of a cloud, we say that the cloud has its highest density where the dots are closest together. Obviously, a particle cannot be "smeared out" over space, but it is, nonetheless, a useful concept. In fact,

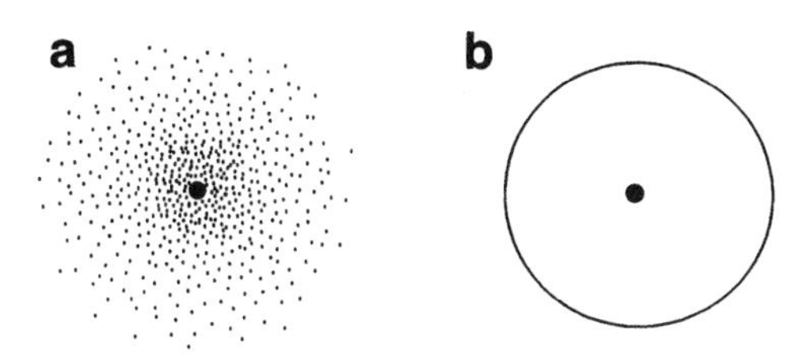

Figure 4.2 ▶ (a) Instantaneous positions of an electron in the $1s$ orbital and (b) the contour surrounding the electron 95% of the time.

one qualitative definition of a covalent bond is the increased probability of finding electrons between two atoms or the increase in density of the electron cloud between two atoms. While the charge cloud does not represent the nature of particles, it provides a way to pictorially describe a probability.

Having plotted the position of the electron in the ground state of a hydrogen atom, we could draw a surface to encompass the positions where the electron was found a specified percentage (perhaps 95%) of the time. In the case of the electron in a $1s$ orbital, the surface is a sphere. Therefore, we say that the $1s$ orbital is *spherical* or has *spherical symmetry* (as shown in Figure 4.2), and the probability of finding the electron depends on r. The quantum state in which the electron resides is referred to as an *orbital*, with this word in no way indicating a *path* of the electron.

Finally, as we have already discussed, the probability of finding the electron in terms of three dimensions (its radial density) is given by

$$P(r)\,dr = [R(r)]^2 4\pi r^2\,dr. \tag{4.35}$$

Figure 2.1 shows a plot of the radial density that indicates that the distance of highest probability (most probable distance) is a_0. This can be shown mathematically as follows. After squaring the radial wave function, we have

$$P(r) = (4\pi)\left(\frac{1}{a_0}\right)^3 r^2 \mathrm{e}^{-2r/a_0}. \tag{4.36}$$

Differentiating with respect to r and setting the derivative equal to 0 gives

$$\frac{dP(r)}{dr} = 0 = (4\pi)\,2r\left(\frac{1}{a_0}\right)^3 \mathrm{e}^{-2r/a_0} - (4\pi)\frac{2}{a_0}r^2\left(\frac{1}{a_0}\right)^3 \mathrm{e}^{-2r/a_0}. \tag{4.37}$$

Therefore,

$$(4\pi)\,2r\left(\frac{1}{a_0}\right)^3 \mathrm{e}^{-2r/a_0} = (4\pi)\frac{2}{a_0}r^2\left(\frac{1}{a_0}\right)^3 \mathrm{e}^{-2r/a_0}, \tag{4.38}$$

from which we find that

$$1 = \frac{r}{a_0}, \tag{4.39}$$

showing that the *most probable radius* is a_0.

To this point, we have dealt with properties of the $1s$ wave function. If we turn to the $2s$ wave function, we find that the radial density plot has a greatly different appearance. Using the $2s$ wave function,

$$\psi_{2s} = \frac{1}{4\sqrt{2\pi}\,a_0^{3/2}}\left(2 - \frac{r}{a_0}\right)\mathrm{e}^{-r/2a_0}, \tag{4.40}$$

the probability of finding the electron has a node where the probability goes to 0 at $r = 2a_0$, which can be seen from the $2 - (r/a_0)$ part of the wave function. The $3s$ wave function is

$$\psi_{3s} = \frac{1}{81\sqrt{3\pi}\,a_0^{3/2}}\left(27 - 18\frac{r}{a_0} + 2\frac{r^2}{a_0^2}\right)e^{-r/3a_0}. \tag{4.41}$$

Since the only part of this wave function (and hence the square of ψ) that can go to 0 is the polynomial, it should be clear that at some value of r (in units of a_0)

$$27 - 18x + 2x^2 = 0, \tag{4.42}$$

where $x = r/a_0$. We can easily solve this quadratic equation to find

$$x = \frac{18 \pm \sqrt{18^2 - 4\,(2)\,(27)}}{4} = 7.10 \text{ and } 1.90. \tag{4.43}$$

Therefore, the probability of finding the electron as a function of distance in the $3s$ state goes to 0 at $r = 1.90a_0$ and $r = 7.10a_0$, so the $3s$ wave function has two nodes. In fact, it is easy to see that the probability has $n - 1$ nodes, where n is the principal quantum number. Figure 4.3 shows the probability density as a function of r for the $2s$ and $3s$ states.

We have already explained that the s wave functions give rise to a spherical surface that encompasses the electron some arbitrary percentage of the time (perhaps 90 or 95%). The surfaces depicting the probabilities of finding the electron in the $2s$ and $3s$ states are spherical, although they are larger than the $1s$ surface. Consequently, the electron density is more diffuse when the electron is in one of these states. We will return to this point in Chapter 9 when we consider bonds formed by the overlap of s orbitals. Surfaces that correspond to the electron density in other states must also be shown.

The surfaces for the p orbitals are shown in Figure 4.4. The appropriate mathematical signs of the wave functions are shown in the figures, and each p orbital has two lobes separated by a nodal plane.

If we consider the cases arising for $l = 2$, we find that there are five orbitals of the d type. These are shown in Figure 4.5. However, if we rotate the d_{xy}, d_{yz}, and d_{xz}, orbitals by 45°, we generate three new orbitals, $d_{x^2-y^2}$, $d_{y^2-z^2}$, and $d_{z^2-x^2}$, having lobes lying along the axes. It can be shown that upon combining these wave functions,

$$\psi_{d_{x^2-y^2}} + \psi_{d_{y^2-z^2}} + \psi_{d_{z^2-x^2}} = 0. \tag{4.44}$$

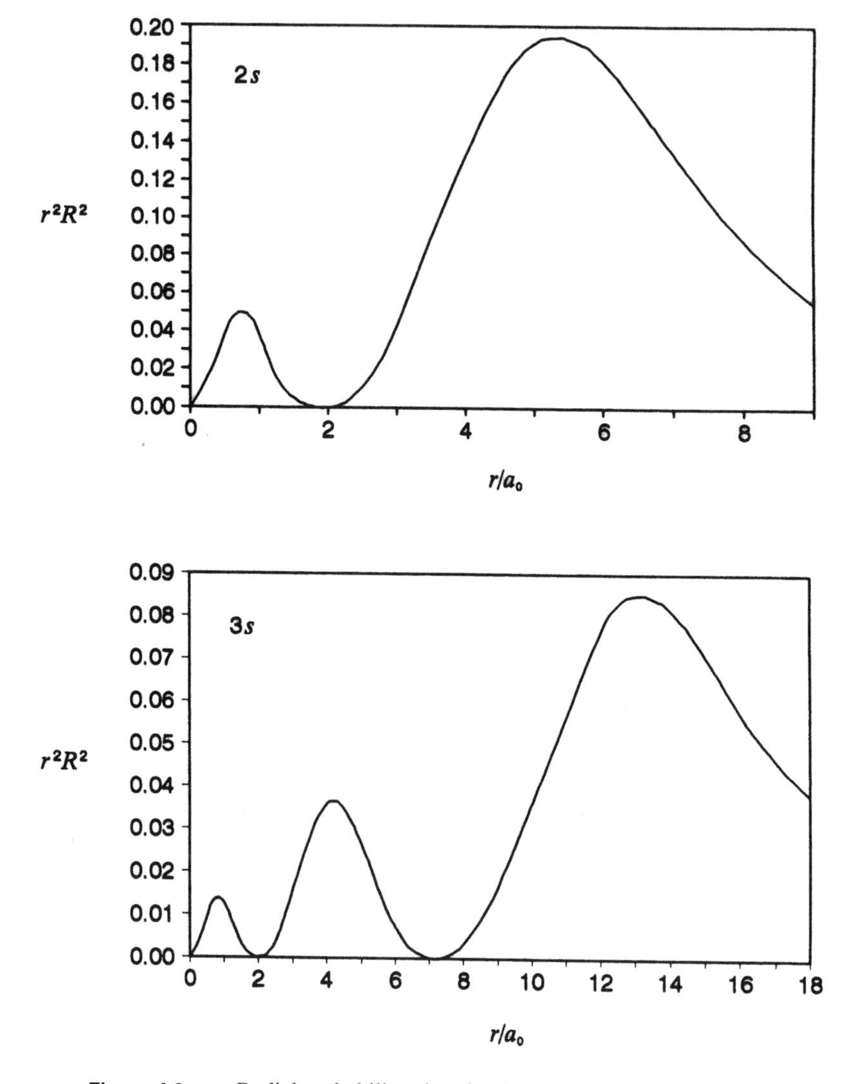

Figure 4.3 ▶ Radial probability plots for the $2s$ and $3s$ wave functions.

Therefore, only two of the three orbitals are independent. In the usual case, we choose $d_{x^2-y^2}$ to be one of the independent orbitals and show the other two combined as the d_{z^2} orbital as shown in Figure 4.5(e). The d_{z^2} orbital is usually shown as a combination of the other two functions, $d_{y^2-z^2}$ and $d_{z^2-x^2}$. For our purposes, the usual diagrams shown in Figure 4.5 will suffice.

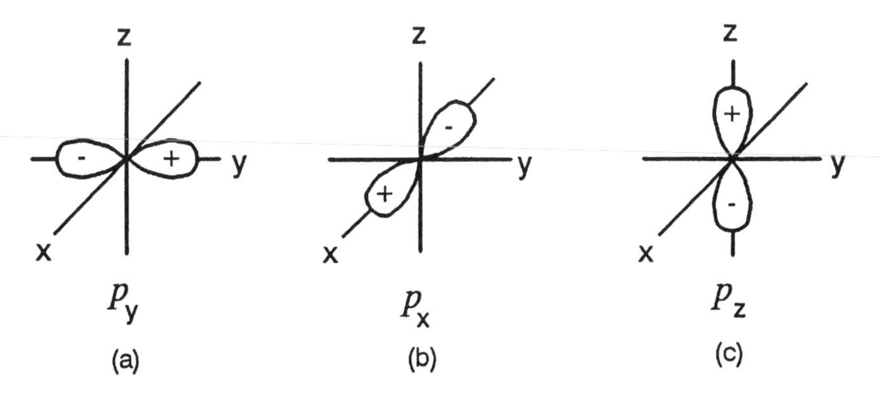

Figure 4.4 ▸ A set of three *p* orbitals.

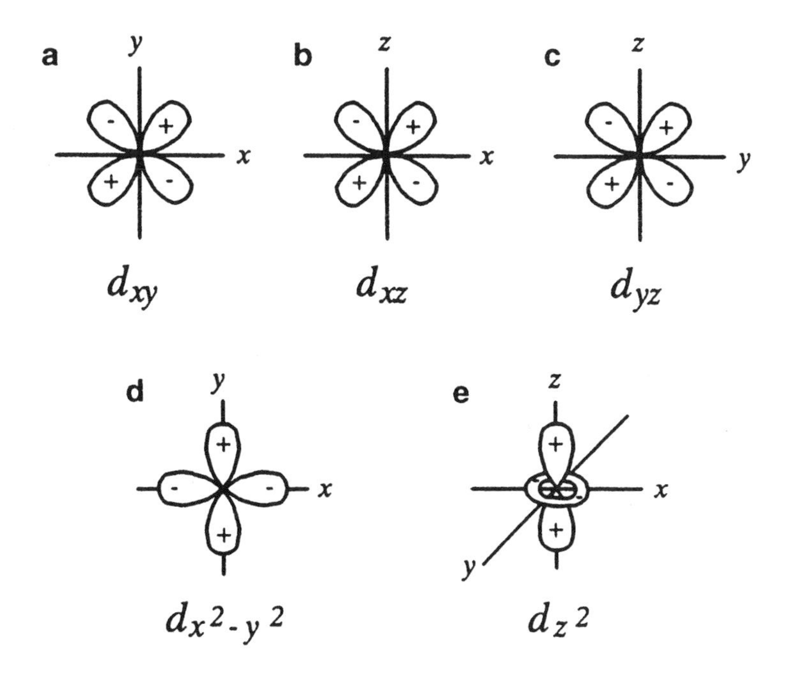

Figure 4.5 ▸ A set of five *d* orbitals.

4.3 Orthogonality

Our ideas about bonding between atoms focus on the combination of atomic orbitals (wave functions). One of the important aspects of orbital combination is that of *orthogonality*. For certain combinations of wave functions, it is found that

$$\int \psi_1 \psi_2 \, d\tau = 0. \tag{4.45}$$

This type of integral, known as an *overlap integral*, gives a measure of the extent to which orbitals overlap in molecules (see Chapter 9). If this condition is met, the wave functions are said to be orthogonal. This relationship shows that there is no effective overlap or congruency of the functions. For example, the p_z and p_x orbitals are perpendicular to each other and the overlap is 0. We can see this pictorially in Figure 4.6. It can also be shown mathematically as follows. The corresponding wave functions are

$$p_z = \frac{1}{4\sqrt{2\pi}} \left(\frac{1}{a_0}\right)^{5/2} r\mathrm{e}^{-r/2a_0} \cos\theta \tag{4.46}$$

$$p_x = \frac{1}{4\sqrt{2\pi}} \left(\frac{1}{a_0}\right)^{5/2} r\mathrm{e}^{-r/2a_0} \sin\theta \cos\phi. \tag{4.47}$$

Therefore,

$$\int \psi_{p_z} \psi_{p_x} \, d\tau \tag{4.48}$$

$$= \frac{1}{32\pi a_0^5} \int_0^\infty \int_0^\pi \int_0^{2\pi} r^4 \mathrm{e}^{-r/a_0} \sin^2\theta \cos\theta \cos\phi \, dr \, d\theta \, d\phi$$

$$= \frac{1}{32\pi a_0^5} \int_0^\infty \int_0^\pi \int_0^{2\pi} r^4 \mathrm{e}^{-r/a_0} \sin^2\theta \cos\theta \, d\theta \int_0^{2\pi} \cos\phi \, d\phi.$$

The exponential integral is of a familiar form and can be evaluated immediately to give

$$\int_0^\infty r^4 \mathrm{e}^{-r/a_0} \, dr = \frac{4!}{(1/a_0)^5}. \tag{4.49}$$

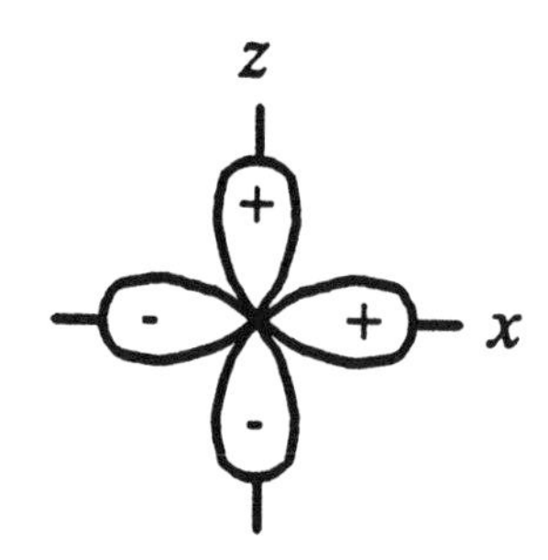

Figure 4.6 ▶ The p_x and p_z orbitals.

The two integrals giving the angular dependence can be looked up in a table of integrals since they are standard forms:

$$\int \cos ax\, dx = \frac{1}{a} \sin ax \tag{4.50}$$

$$\int \sin^n ax \cos ax\, dx = \frac{1}{a\,(n+1)} \sin^{n+1} ax. \tag{4.51}$$

For the integrals that we are using, $a = 1$ and $n = 2$, so

$$\int_0^{2\pi} \cos\phi\, d\phi = \sin\phi|_0^{2\pi} = \sin 2\pi - \sin 0 = 0 \tag{4.52}$$

$$\int_0^{\pi} \sin^2\theta \cos\theta\, d\theta = \frac{1}{3} \sin^3\phi|_0^{\pi} = \frac{1}{3}\left[\sin^3\pi - \sin^3 0\right] = 0 \tag{4.53}$$

Therefore, $\int \psi_{p_z}\psi_{p_x}\, d\tau = 0$, so the orbitals are orthogonal, as expected. Note that it is the angular dependence being different for the two orbitals that leads to orthogonality.

The implications of the orthogonality of orbitals are of great importance. For example, we can immediately see that certain types of orbital overlap will not occur; two of these are shown in Figure 4.7. For a given atom it can be shown that if ψ_1 and ψ_2 are orbitals that have different symmetry types, they must be orthogonal. Within the same *type* of orbital, p_x, p_y, and p_z, are orthogonal, as are the five orbitals in a set of d orbitals. However, we can also see that numerous types of overlap, some of which are shown in Figure 4.8, lead to favorable situations (overlap integral > 0). We will return to the discussion of the overlap of atomic orbitals in Chapters 9 and 11.

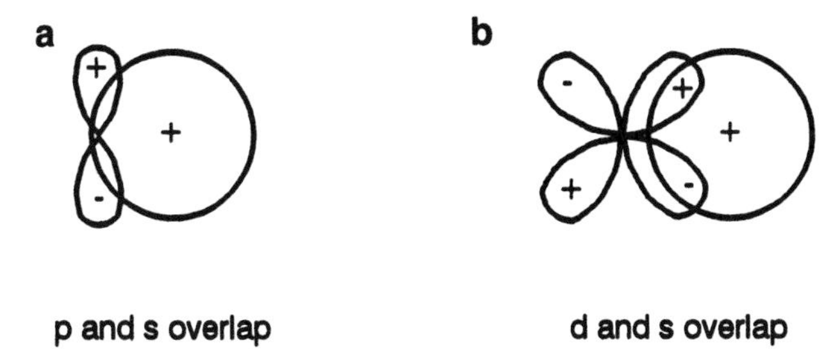

Figure 4.7 ▸ Interaction of orbitals giving no overlap (a) p and s overlap. (b) d and s overlap.

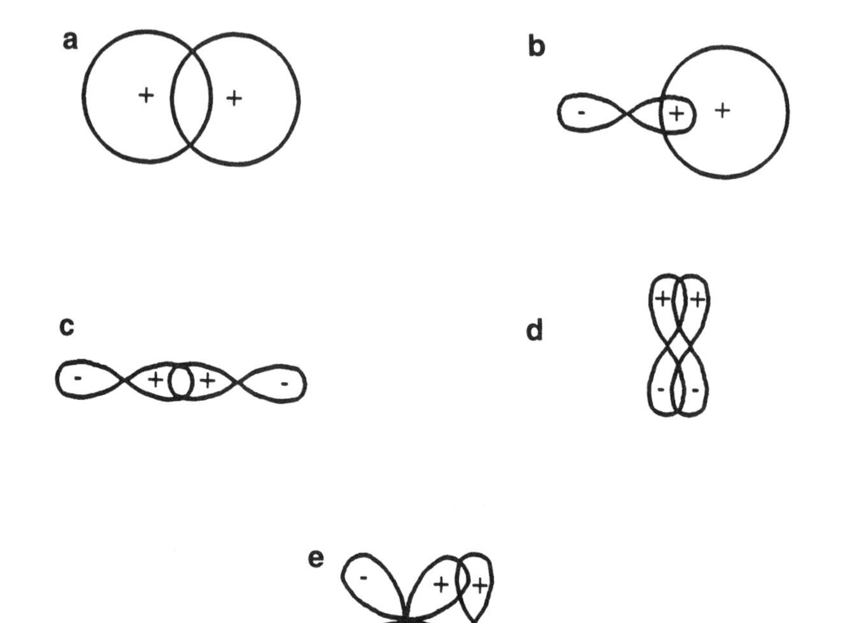

Figure 4.8 ▶ Favorable overlap of orbitals.

4.4 Approximate Wave Functions and the Variation Method

For many problems, it is not practical to obtain a wave function by the exact solution of a wave equation that describes the system. It is still possible to perform many types of calculations, and one of the most useful techniques is that known as the *variation method*. We will now illustrate the use of that technique by considering a simple problem. We will also make use of it in the next chapter in dealing with the problem of the helium atom and in Chapter 9 when dealing with diatomic molecules. Using the procedure introduced in Section 2.4, the expectation value for the energy is given by

$$E = \frac{\int \psi^* \hat{H} \psi \, d\tau}{\int \psi^* \psi \, d\tau}. \tag{4.54}$$

If the exact form of the wave function is unknown, we begin by assuming some form of the wave function. It is frequently described as "guessing" a *trial wave function*, but this is hardly the case. We already know the general form of the wave functions found from the solution of the hydrogen atom problem so a function of that form is a good starting point.

The variation theorem provides the basis for the variation method. This theorem states that the correct energy is obtained from the previous equation only when the correct wave function is used. Any "incorrect" wave function will give an energy that is higher than the actual energy. We will accept this premise without providing a proof at this point. If we choose a trial wave function, ψ_i, the energy calculated using it, E_i, is greater than the correct energy, E_0. That is, for any incorrect wave function, $E_i > E_0$. We choose to try a wave function that has some adjustable parameter so that we can vary its value to "improve" the wave function. This "improved" wave function can then be used to calculate an improved value for the energy, etc.

A simple illustration of the variation method is provided by the hydrogen atom in the 1s state. Let us assume a form of the trial wave function

$$\psi = \mathrm{e}^{-br}, \tag{4.55}$$

where b is a constant whose actual value can be changed as we gain information about it. For the hydrogen atom, $V = -e^2/r$, so the Hamiltonian operator is

$$\hat{H} = -\frac{h^2}{8\pi^2 m}\nabla^2 - \frac{e^2}{r}. \tag{4.56}$$

The energy depends only on r for the 1s state of the hydrogen atom so the angular portion of the Laplacian can be omitted and replaced by the factor 4π after integration. Therefore, we will use the radial portion of ∇^2,

$$\nabla^2 = \frac{1}{r^2}\frac{\partial}{\partial r}r^2\frac{\partial}{\partial r}. \tag{4.57}$$

When we operate on the trial wave function, ψ, with this operator,

$$\nabla^2\psi = \frac{1}{r^2}\frac{\partial}{\partial r}r^2\frac{\partial}{\partial r}\mathrm{e}^{-br} \tag{4.58}$$

Taking the derivatives and simplifying gives

$$\nabla^2\psi = \left(b^2 - \frac{2b}{r}\right)\mathrm{e}^{-br}. \tag{4.59}$$

When this result is substituted into Eq. (4.54), we obtain

$$E = \frac{\int_0^\infty \mathrm{e}^{-br}\left(-\frac{h^2}{8\pi^2 m}\nabla^2 - \frac{e^2}{r}\right)\mathrm{e}^{-br}4\pi r^2\,dr}{\int_0^\infty \left(\mathrm{e}^{-br}\right)\left(\mathrm{e}^{-br}\right)4\pi r^2\,dr} \tag{4.60}$$

or, after substituting for ∇^2 and removing the factor of 4π and canceling,

$$E = \frac{\int_0^\infty \mathrm{e}^{-br}\left(-\frac{h^2}{8\pi^2 m}\left(b^2 - \frac{2b}{r}\right)\mathrm{e}^{-br} - \frac{e^2}{r}\mathrm{e}^{-br}\right) r^2\,dr}{\int_0^\infty \mathrm{e}^{-2br} r^2\,dr}. \tag{4.61}$$

Expanding the terms in the numerator by multiplying, we obtain

$$E = \frac{\int_0^\infty -\frac{h^2 b^2}{8\pi^2 m} r^2 \mathrm{e}^{-2br}\,dr + \int_0^\infty \frac{2h^2 b}{8\pi^2 m} r\mathrm{e}^{-2br}\,dr - \int_0^\infty e^2 r\mathrm{e}^{-2br}\,dr}{\int_0^\infty \mathrm{e}^{-2br} r^2\,dr}. \tag{4.62}$$

Fortunately, each of these integrals is of the easily recognized form

$$\int_0^\infty x^n \mathrm{e}^{-ax}\,dx = \frac{n!}{a^{n+1}}. \tag{4.63}$$

Therefore, evaluating the integrals gives

$$E = \frac{-\frac{h^2 b^2}{8\pi^2 m}\frac{2}{8b^3} + \frac{h^2 b}{4\pi^2 m}\frac{1}{b^2} - e^2\frac{1}{4b^2}}{\frac{2}{8b^3}}. \tag{4.64}$$

Finally, after simplification of this expression, we obtain

$$E = \frac{h^2 b^2}{8\pi^2 m} - be^2. \tag{4.65}$$

This equation gives the energy in terms of fundamental constants and the adjustable parameter, b. It is necessary to find the value of b that will give the minimum energy. This is done by taking the derivative of E with respect to b and setting the derivative equal to 0. We thus obtain

$$\frac{\partial E}{\partial b} = \frac{2h^2 b}{8\pi^2 m} - e^2 = 0. \tag{4.66}$$

Solving this equation for b gives

$$b = \frac{4\pi^2 m e^2}{h^2}. \tag{4.67}$$

When we substitute this value for b into Eq. (4.65) for the energy, we obtain

$$E = -\frac{2\pi^2 m e^4}{h^2}. \tag{4.68}$$

This expression is exactly the same as that found using the Bohr model (see Chapter 1)! How were we so lucky as to obtain the correct energy in a single "improvement" of the wave function? The answer is that we "guessed" the correct form of the wave function to use as the trial one. We knew that an exponential involving r was the form of the actual wave function so the variation method enabled us to evaluate the constant in a single calculational cycle. The variation method will be used as the basis for other types of calculations in later chapters.

References for Further Reading

- Alberty, R. A., and R. J. Silbey (1996). *Physical Chemistry*, 2nd ed. Wiley, New York. One of the best introductions to quantum mechanics of atoms available.
- Eyring, H., Walter, J., and Kimball, G. E. (1944). *Quantum Chemistry*. Wiley, New York. Excellent mathematical treatment of the hydrogen atom problem.
- Herzberg, G. (1944) *Atomic Spectra and Atomic Structure*. Dover, New York. A good reference for line spectra and related topics.
- Laidler, K. J., and Meiser, J. H. (1982). *Physical Chemistry*, Chap. 11. Benjamin–Cummings, Menlo Park, CA. A clear introduction to atomic structure.
- Pauling, L., and Wilson, E. B. (1935). *Introduction to Quantum Mechanics*. McGraw–Hill, New York. A standard reference that has stood the test of time.

Problems

1. Calculate the velocity of an electron in the $n = 1$ state of a hydrogen atom.

2. Show that the de Broglie wavelength of an electron moving at the velocity found in Problem 1 corresponds to the circumference of the first Bohr radius.

3. What is the total number of electrons that can be accommodated if $n = 5$?

4. The potential energy for an electron attracted to a $+1$ nucleus is $V = -e^2/r$. Using the variation method, determine $\langle V \rangle$ in this case.

5. Determine the value for $\langle r \rangle$ for an electron in the $2p_z$ state of the hydrogen atom.

6. Use the procedure described in the text to determine the probability that the electron in the $1s$ state of hydrogen will be found outside a_0.

7. Use the procedure described in the text to determine the probability that the electron in a hydrogen atom will be found between a_0 and $2a_0$.

8. Calculate the average or expectation energy $\langle E \rangle$ for the electron in the $1s$ state of the hydrogen atom.

9. We have shown that the most probable radius of the hydrogen atom is $(3/2)a_0$. Consider the hydrogen atom as a particle in a one-dimensional box (the electron can travel on either side of the nucleus) and calculate the energy (in eV) of the electron in the state with $n = 1$.

10. If a sphere is to be drawn with the nucleus at the center, how large must the sphere be to encompass the electron in a hydrogen atom 90% of the time for the $n = 1$ state?

11. Show that the wave functions for the $1s$ and $2s$ states in the hydrogen atom are orthogonal.

▶ Chapter 5

More Complex Atoms

As we will see, wave equations cannot be solved exactly for complex atoms. Although a rigorous quantum mechanical treatment cannot result in a closed form solution to the wave equation for a complex atom, such solutions are usually not necessary for an understanding of most aspects of chemical bonding. At this point, we present the approaches used to describe the helium atom as well as some of the empirical and experimental properties of atoms.

5.1 The Helium Atom

Although it is easy to formulate the wave equation for atoms that are more complex than hydrogen, such equations cannot be solved exactly. We shall consider the helium atom to see why this is so. The helium atom can be represented as shown in Figure 5.1. Taking into account the motion of the electrons and the electrostatic interaction the Hamiltonian operator can be written as

$$\hat{H} = -\frac{\hbar^2}{2m}\nabla_1^2 - \frac{\hbar^2}{2m}\nabla_2^2 - \frac{2e^2}{r_1} - \frac{2e^2}{r_2} + \frac{e^2}{r_{12}}, \tag{5.1}$$

which leads directly to the wave equation

$$\left(-\frac{\hbar^2}{2m}\nabla_1^2 - \frac{\hbar^2}{2m}\nabla_2^2 - \frac{2e^2}{r_1} - \frac{2e^2}{r_2} + \frac{e^2}{r_{12}}\right)\psi = E\psi. \tag{5.2}$$

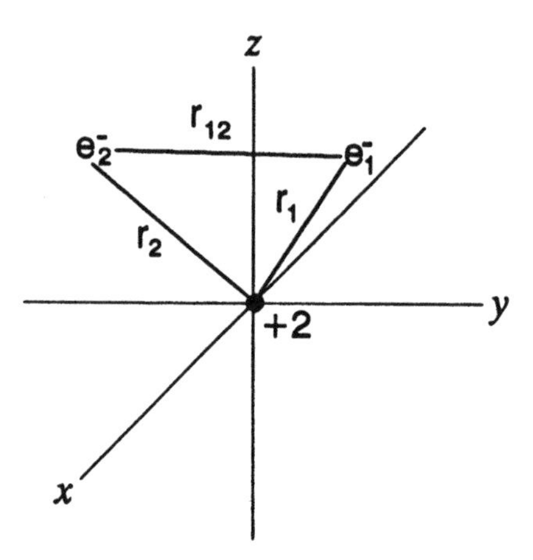

Figure 5.1 ▶ The helium atom coordinates.

For the hydrogen atom, the Hamiltonian operator has only one term involving $1/r$ (where r is the distance of the electron from the nucleus), and that difficulty is avoided by the use of polar coordinates. In the case of the helium atom, even a change to polar coordinates does not help because of the term containing $1/r_{12}$. Thus, when the term involving $1/r_{12}$ is included, the variables cannot be separated.

One approach to solving the wave equation for the helium atom is by constructing a trial wave function and using the variation method to optimize it. For a trial wave function we will take

$$\psi = \phi_1\phi_2 = \frac{Z'^3}{\pi a_0^3} e^{-Z'r_1/a_0} e^{-Z'r_2/a_0}, \tag{5.3}$$

where ϕ_1 and ϕ_2 are hydrogen-like wave functions and Z' is an *effective* nuclear charge that is less than the actual value of 2. The Hamiltonian can be written as

$$\hat{H} = -\frac{\hbar^2}{2m}(\nabla_1^2 + \nabla_2^2) - Ze^2\left(\frac{1}{r_1} + \frac{1}{r_2}\right) + \frac{e^2}{r_{12}} \tag{5.4}$$

so that

$$\hat{H}\psi = E\psi = \hat{H}\phi_1\phi_2 = E_{\mathrm{H}}\phi_1\phi_2, \tag{5.5}$$

where E_{H} is the energy of the hydrogen atom in the $1s$ state. The quantity $Z'E_{\mathrm{H}}$ is the energy that results when the Bohr model is applied to hydrogen-like species where the nuclear charge is not 1. Therefore,

$$\hat{H}\phi_1 = -\frac{\hbar^2}{2m}\nabla^2\phi_1 = \frac{Z'e^2\phi_1}{r_1} - Z'^2 E_{\mathrm{H}}\phi_1. \tag{5.6}$$

A similar equation can be written using the second atomic wave function, ϕ_2. Therefore,

$$\begin{aligned} \hat{H}\phi_1\phi_2 &= \frac{Z'e^2}{r^2}\phi_1\phi_2 - Z'^2 E_{\mathrm{H}}\phi_1\phi_2 \\ &\quad + \frac{Z'e^2}{r_1}\phi_1\phi_2 - Z'^2 E_{\mathrm{H}}\phi_1\phi_2 \\ &\quad - Z'e^2\left(\frac{1}{r_1}+\frac{1}{r_2}\right)\phi_1\phi_2 + \frac{e^2}{r_{12}}\phi_1\phi_2. \end{aligned} \tag{5.7}$$

We can now multiply by $\phi_1^*\phi_2^*$ to obtain

$$\begin{aligned} \phi_1^*\phi_2^*\hat{H}\phi_1\phi_2 &= \phi_1^*\phi_2^*\frac{Z'e^2}{r_1}\phi_1\phi_2 + \phi_1^*\phi_2^*\frac{Z'e^2}{r_2}\phi_1\phi_2 \\ &\quad - \phi_1^*\phi_2^* 2Z'^2 E_{\mathrm{H}}\phi_1\phi_2 - \phi_1^*\phi_2^* Ze^2\left(\frac{1}{r_1}+\frac{1}{r_2}\right)\phi_1\phi_2 \\ &\quad + \phi_1^*\phi_2^*\frac{e^2}{r_{12}}\phi_1\phi_2. \end{aligned} \tag{5.8}$$

After writing the terms as integrals, the total energy can be expressed as

$$\begin{aligned} E &= -2Z'^2 E_{\mathrm{H}} + (Z'-Z)\,e^2 \int \phi_1^*\phi_2^*\left(\frac{1}{r_1}+\frac{1}{r_2}\right)\phi_1\phi_2\,d\tau \\ &\quad + \int \phi_1^*\phi_2^*\frac{e^2}{r_{12}}\phi_1\phi_2\,d\tau. \end{aligned} \tag{5.9}$$

The middle term on the right-hand side of this equation is

$$\int \phi_1^*\frac{1}{r_1}\phi_1\,d\tau_1 \int \phi_2^*\phi_2\,d\tau_2 + \int \phi_2^*\frac{1}{r_2}\phi_2\,d\tau_2 \int \phi_1^*\phi_1\,d\tau_1. \tag{5.10}$$

Two of the integrals can be set equal to 1,

$$\int \phi_2^*\phi_2\,d\tau_2 = \int \phi_1^*\phi_1\,d\tau_1 = 1. \tag{5.11}$$

Therefore, the middle term on the right-hand side of Eq. (5.9) reduces to

$$(Z'-Z)\,e^2 2\int \phi_1^*\frac{1}{r_1}\phi_1\,d\tau_1 = 4\,(Z'-Z)\,Z'E_{\mathrm{H}}. \tag{5.12}$$

The energy can now be written from Eqs. (5.9) and (5.12) as

$$E = -2Z'^2 E_{\mathrm{H}} + 4\,(Z'-Z)\,Z'E_{\mathrm{H}} + \frac{5}{4}Z'E_{\mathrm{H}} \tag{5.13}$$

$$E = -2Z'^2 E_{\mathrm{H}} + 4Z'^2 E_{\mathrm{H}} - 4ZZ'E_{\mathrm{H}} + \frac{5}{4}Z'E_{\mathrm{H}}. \tag{5.14}$$

Having obtained an expression for the energy in terms of the effective nuclear charge, Z', we now wish to determine the value of Z' that results in the minimum energy, which is done by taking the derivative and setting it equal to 0,

$$\left(\frac{\partial E}{\partial Z'}\right) = 0 = -4Z'E_{\mathrm{H}} + 8Z'E_{\mathrm{H}} - 4ZE_{\mathrm{H}} + \frac{5}{4}E_{\mathrm{H}}. \tag{5.15}$$

Solving this expression for Z', we obtain

$$Z' = Z - \frac{5}{16}. \tag{5.16}$$

Thus, each electron experiences an effective nuclear charge of $27/16$ instead of the nuclear charge of 2 because of screening by the other electron.

Having calculated that the effective nuclear charge in the helium atom is $27/16$, we can now use that value in Eq. (5.14) to determine the total binding energy for the two electrons. When the substitution of the value for Z' is made, we find that $E = +5.696$ eV and since the ionization potential for the hydrogen atom, E_{H}, is 13.6 eV, the total ionization potential for the two electrons in helium is calculated to be 77.5 eV. The experimental value is approximately 79.0 eV so the calculation using the variation method yields a value that is in reasonable agreement with the observed value.

The discussion of the variation method presented in Section 4.4 and its application to the calculation of the total electron binding energy in the helium atom have shown that the variation method is an important tool in quantum mechanics. Another such tool is the *perturbation method*. The basic idea behind perturbation theory is that the system does not behave perfectly because of some "slight" deviation from a system that can be treated exactly. In the orbits of planets, deviation from perfect orbits result from gravitational forces, which become more important as the planets get closer together in their orbits. In the harmonic oscillator model, the perturbation might be a potential that is not expressed exactly by $\frac{1}{2}kx^2$ (see Section 6.4). From Eq. (5.2), we see that the helium atom could be treated as the sum of two hydrogen atomic problems if the e^2/r_{12} term that arises from repulsion of the electrons were not present in the Hamiltonian. Therefore, the repulsion term is treated as a "slight" perturbation of an otherwise "perfect" system that could be solved exactly. We will present an overview of the general principles to show how the method is approached then show how it applies to the helium atom case.

For an unperturbed system that can be treated exactly, the wave equation can be written as

$$\hat{H}^{\mathrm{o}}\psi^{\mathrm{o}} = E^{\mathrm{o}}\psi^{\mathrm{o}}. \tag{5.17}$$

The solutions are the wave functions $\psi_0^{\text{o}}, \psi_1^{\text{o}}, \psi_2^{\text{o}}, \ldots$ and the energy eigenvalues are $E_0^{\text{o}}, E_1^{\text{o}}, E_2^{\text{o}}, \ldots,$ etc. The Hamiltonian is that which is appropriate for the particular system, but it has the form

$$\hat{H} = -\frac{\hbar^2}{2m}\nabla^2 + V. \tag{5.18}$$

If the system becomes slightly perturbed, that perturbation is represented by a slight alteration in the potential term of the Hamiltonian. Therefore, we can write the Hamiltonian as

$$\hat{H} = \hat{H}^{\text{o}} + \lambda\hat{H}' + \lambda^2\hat{H}' + \cdots, \tag{5.19}$$

where λ is a parameter that gives the extent of perturbation and $\hat{H}'$ is the adjustment to the Hamiltonian. In performing the calculation of the first-order perturbation, the terms beyond $\hat{H}'$ in the series are ignored.

Because the perturbation energy is presumed to be small compared to the total energy, the wave function for the kth state of the system is written as

$$\psi_k = \psi_k^{\text{o}} + \lambda\psi_k', \tag{5.20}$$

and the corresponding energy of the state is given by

$$E_k = E_k^{\text{o}} + \lambda E_k'. \tag{5.21}$$

Using the expressions above and the fundamental relationship that $\hat{H}\psi = E\psi$, the first-order correction is obtained by omitting terms in λ^2 (since λ is small). Therefore, for the unperturbed kth state we obtain the relationship

$$\hat{H}^{\text{o}}\psi_k^{\text{o}} = E^{\text{o}}\psi_k^{\text{o}}. \tag{5.22}$$

For the first-order perturbation,

$$\hat{H}^{\text{o}}\psi_k' + \hat{H}'\psi_k^{\text{o}} - E_k^{\text{o}}\psi_k' - E_k'\psi_k^{\text{o}} = 0 \tag{5.23}$$

can also be written as

$$(\hat{H}^{\text{o}} - E_k^{\text{o}})\psi_k' = (E_k' - \hat{H}')\psi_k^{\text{o}}. \tag{5.24}$$

A linear combination of solutions is used to represent the wave function of the perturbed kth state in terms of the wave functions for all of the i unperturbed states. This series is written as

$$\psi_k' = \sum_i a_i\psi_i^{\text{o}}. \tag{5.25}$$

From a combination of Eq. (5.22) and Eq. (5.25), we obtain

$$\hat{H}^{\mathrm{o}}\psi_k' = \sum_i a_i \hat{H}^{\mathrm{o}}\psi_i^{\mathrm{o}} = \sum_i a_i E_i^{\mathrm{o}}\psi_i^{\mathrm{o}}, \tag{5.26}$$

which can be written as

$$\sum_i (a_i E_i^{\mathrm{o}}\psi_i^{\mathrm{o}} - a_i E_k^{\mathrm{o}}\psi_i^{\mathrm{o}}) = E_k'\psi_k^{\mathrm{o}} - \hat{H}'\psi_k^{\mathrm{o}} \tag{5.27}$$

or

$$\sum_i a_i (E_i^{\mathrm{o}} - E_k^{\mathrm{o}})\psi_i^{\mathrm{o}} = E_k'\psi_k^{\mathrm{o}} - \hat{H}'\psi_k^{\mathrm{o}}. \tag{5.28}$$

Multiplying both sides of this equation by $\psi_k^{\mathrm{o}*}$ (see Section 2.4) and integrating over all space, the left-hand side of Eq. (5.28) can be written as

$$\int \psi_k^{\mathrm{o}} \sum a_i (E_i^{\mathrm{o}} - E_k^{\mathrm{o}})\psi_i^{\mathrm{o}}\, d\tau. \tag{5.29}$$

This expression can be greatly simplified since for orthogonal wave functions the integral

$$\int \psi_i^{\mathrm{o}*}\psi_k^{\mathrm{o}}\, d\tau = 0 \tag{5.30}$$

gives a value of 0 when $i \neq k$ and a value of 1 when $i = k$. When $i = k$, we see that $E_k^{\mathrm{o}} = E_i^{\mathrm{o}}$ so the entire integral vanishes and the left-hand side of Eq. (5.28) must equal 0. We know that the right-hand side of Eq. (5.28) must equal 0, so

$$\int \psi_k^{\mathrm{o}*}(E' - \hat{H}')\psi_k^{\mathrm{o}}\, d\tau = 0. \tag{5.31}$$

Separation of the integral is possible, which allows us to write (the perturbation energy, E', is a constant and is removed from the integral)

$$E' \int \psi_k^{\mathrm{o}*}\psi_k^{\mathrm{o}}\, d\tau - \int \psi_k^{\mathrm{o}*}\hat{H}'\psi_k^{\mathrm{o}}\, d\tau = 0. \tag{5.32}$$

For normalized wave functions, the integral multiplied by E' is equal to 1. Therefore,

$$E' = \int \psi_k^{\mathrm{o}*}\hat{H}'\psi_k^{\mathrm{o}}\, d\tau. \tag{5.33}$$

This result shows that the perturbation energy correction to the kth state is the familiar expectation value with only the perturbation Hamiltonian being used in the integration. After E' is calculated, the total energy of the kth energy level (the $1s$ state in helium in this case) will be given by $E^{\text{o}} + E'$.

Application of the perturbation method to the helium atom involves treating the term e^2/r_{12} as the perturbation to an otherwise exactly solvable system consisting of two hydrogen atoms. We already know that the ionization potential of the hydrogen atom is 13.6 eV. From the exact solution of the approximate equation obtained by neglecting the e^2/r_{12} term in the Hamiltonian, the energy for the $1s$ state in helium is

$$E^{\text{o}} = -Z^2 E_{\text{H}} - Z^2 E_{\text{H}} = -2Z^2 E_{\text{H}}. \tag{5.34}$$

The perturbation term involving e^2/r_{12} gives an energy E', which can be expressed as

$$E' = \iint \psi^*_{(1)} \psi^*_{(2)} (e^2/r_{12}) \psi_{(1)} \psi_{(2)}\, d\tau_1\, d\tau_2, \tag{5.35}$$

where e^2/r_{12} is the perturbation operator, $\hat{H}'$. Of course for the $1s$ wave function, $\psi^* = \psi$. This integral must be evaluated to give the perturbation correction to the energy $-2Z^2 E_{\text{H}}$, which was obtained by neglecting the repulsion between the two electrons. Each electron is represented as a spherically symmetric charge field and the integral representing their interaction can be transformed to give

$$\frac{Z^6}{\pi^2 a_0^6} \int \frac{\exp(-2Zr_1/a_{\text{o}}) \exp(-2Zr_2/a_{\text{o}})}{r_{12}} dV.$$

In this integral, the exponential functions are charge distributions of two spherically symmetric electrostatic fields produced by the two electrons. Evaluation the integral[1] leads to a perturbation energy of

$$E' = \frac{5}{4} Z E_{\text{H}}. \tag{5.36}$$

Therefore, the total energy for the $1s$ level in the helium atom is

$$E = -2Z^2 E_{\text{H}} + \frac{5}{4} Z E_{\text{H}} = -\left[2Z^2 - \frac{5}{4} Z\right] E_{\text{H}}. \tag{5.37}$$

Note that the perturbation caused by the repulsion of the two electrons raises the energy (destabilization) of the $1s$ level so that the total binding energy of the two electrons is not as great as it would be for a $+2$ nucleus with no

[1] See Chapter 6 of the book by Pauling and Wilson for details.

repulsion between the electrons. Substituting $Z = 2$ and $E_{\mathrm{H}} = 13.6$ eV we find that the *binding* energy is

$$E = -\left[2(2)^2 - 2\left(\frac{5}{4}\right)\right](13.6) = -5.50(13.6) = -74.8 \text{ eV}. \quad (5.38)$$

Therefore, the total *ionization* energy is 74.8 eV. As mentioned earlier, the experimental value is approximately 79.0 eV.

In this section, we have shown the application of two of the very important approximation methods widely used in quantum mechanical calculations. Other applications of the variation method to molecules will be shown in Section 9.3.[2] From the discussion presented in this section, it should be clear why variation and perturbation methods are among the most important techniques used in quantum mechanical calculations.

While we cannot solve exactly the wave equation (for even the helium atom), we can arrive at approximate solutions. In this case, evaluating the total energy by the variation and perturbation methods leads to values that are quite close to the actual binding energy of the two electrons. It is a simple matter to write the wave equation for complex atoms by writing the Hamiltonian in terms of the various attraction and repulsion energies, but various approximation methods must be used to solve the equations. In the next section, we will discuss a different approach to obtaining wave functions for complex atoms.

5.2 Slater Wave Functions

The exact solution of the Schrödinger wave equation for complex atoms is not possible. However, examination of the form of the wave functions obtained for the hydrogen atom suggests that approximate wave functions might be obtained if we were to take into account the mutual electron repulsion. Such a procedure has been devised by J. C. Slater, and the approximate wave functions that result are known as *Slater wave functions* [or Slater-type orbitals (STO)]. The wave functions are written in the form

$$\psi_{n,l,m} = R_{n,l}(r)\, e^{-Zr/a_0 n}\, Y_{l,m}(\theta, \phi). \quad (5.39)$$

Specifically, the wave functions have the form

$$\psi_{n,l,m} = r^{n^*-1} e^{-(Z-s)r/a_0 n^*}\, Y_{l,m}(\theta, \phi), \quad (5.40)$$

[2]For a rigorous presentation of the theory of perturbation and variation methods see the book by Pauling and Wilson cited in the reference list.

where s is the screening constant and n^* is a parameter that varies with the principal quantum number, n. The screening constant, s, for a given electron is determined by considering all of the contributions from all the populated orbitals in the atom. The electrons are grouped according to the procedure that follows, and the weightings from each group are determined according to the other rules:

1. The electrons are grouped in this manner:

$$1s \mid 2s \mid 2p \mid 3s3p \mid 3d \mid 4s4p \mid 4d \mid 4f \mid 5s5p \mid 5d \mid \ldots$$

2. No contribution to the screening of an electron is considered as arising from orbitals *outside* the orbital holding the electron for which the wave function is being written.

3. For the $1s$ level, the contribution is 0.30, but for other groups 0.35 is added for each electron in that group.

4. For an electron in an s or p orbital, 0.85 is added for each other electron when the principal quantum number is one less than that for the orbital being written. For still lower levels, 1.00 is added for each electron.

5. For electrons in d and f orbitals, 1.00 is added for each electron residing below the one for which the wave function is being written.

6. The value of n^* varies with n as follows:

$$\begin{array}{lllllll} n = 1 & 2 & 3 & 4 & 5 & 6 \\ n^* = 1 & 2 & 3 & 3.7 & 4.0 & 4.2. \end{array}$$

Suppose we need to write the Slater wave function for an electron in a $2p$ orbital of oxygen ($Z = 8$). For that electron, $n = 2$ so $n^* = 2$ also. The screening constant for the fourth electron in the $2p$ level is determined as follows: For the two electrons in the $1s$ level, $2(0.85) = 1.70$. For the five electrons in the $2s$ and $2p$ levels, $5(0.35) = 1.75$. Summing these contributions to the screening constant for the electron in question, we find that $s = 3.45$ and the effective nuclear charge is $8 - 3.45 = 4.55$, so that $(Z - s)/n^* = 2.28$. Therefore, the Slater wave function for an electron in the $2p$ level of oxygen can be written as

$$\psi = re^{-2.28r/a_0} Y_{2,m}(\theta, \phi). \tag{5.41}$$

Earlier we used the variation method to determine the optimum value of the nuclear charge for helium. While the actual nuclear charge is 2, the variation method predicts $27/16 = 1.6875$ as the effective nuclear charge that each electron experiences. This difference is an obvious result of screening by the other electron. It is now possible for us to compare the result from the variation method to screening in the helium atom obtained using Slater's rules. For an electron in the 1*s* level, the only screening is that of the other electron, for which the value 0.30 is used. The effective nuclear charge is $(Z - s)/n^* = (2 - 0.30)/1 = 1.70$, in good agreement with the result obtained by the variation method.

Slater-type orbitals are frequently useful as a starting point in other calculations. As we have seen, an approximate wave function can give useful results when used in the variation method. Also, *some* atomic wave function must be used in the construction of wave functions for molecules. Consequently, these approximate, semiempirical wave functions are especially useful and molecular orbital calculations are frequently carried out using a STO basis set. More discussion of this topic will appear in Chapter 12.

5.3 Electron Configurations

As we have seen, four quantum numbers are required to completely describe an electron in an atom. There are certain restrictions on the values these quantum numbers can have. Thus, $n = 1, 2, 3, \ldots$, and $l = 0, 1, 2, \ldots, (n - 1)$. For a given value of n, the quantum number l can have all integer values from 0 to $(n - 1)$. The quantum number m can have the series of values $+1, +(l-1), \ldots, 0, \ldots, -(l-1), -l$. Thus, there are $(2l - 1)$ values for m. The fourth quantum number, s, can have values of $\pm(\frac{1}{2})$, with this being the spin angular momentum in units of $\frac{h}{2\pi}$.

We can write a set of four quantum numbers to describe each electron in an atom. It is necessary to use the *Pauli Exclusion Principle*, which states that no two electrons in the same atom can have the same set of four quantum numbers. In the case of the hydrogen atom, we begin by recognizing that the lower n values represent states of lower energy. For hydrogen, we can write four quantum numbers to describe the electron as

$$\begin{aligned} n &= 1 \\ l &= 0 \\ m &= 0 \\ s &= +\tfrac{1}{2} \quad \left(\text{or} - \tfrac{1}{2}\right). \end{aligned}$$

The value chosen for s is arbitrary. For helium, which has two electrons, we can write

Electron 1	Electron 2
$n = 1$	$n = 1$
$l = 0$	$l = 0$
$m = 0$	$m = 0$
$s = +\frac{1}{2}$	$s = -\frac{1}{2}$.

An atomic energy level is denoted by the n value followed by a letter (s, p, d, or f to denote $l = 0$, 1, 2, or 3, respectively), and the ground state for hydrogen is $1s^1$, while that for helium is $1s^2$. The two sets of quantum numbers written above complete the sets that can be written for the first shell with $n = 1$.

For $n = 2$, l can have the values of 0 and 1. In general, the levels increase in energy as the sum $(n + l)$ increases. Taking the value of $l = 0$ first, the state with $n = 2$ and $l = 0$ is designated as the $2s$ state, and like the $1s$ state it can hold two electrons:

Electron 1	Electron 2
$n = 2$	$n = 2$
$l = 0$	$l = 0$
$m = 0$	$m = 0$
$s = +\frac{1}{2}$	$s = -\frac{1}{2}$.

These two sets of quantum numbers describe the electrons residing in the $2s$ level. Taking now the value $l = 1$, we find that six sets of quantum numbers can be written:

Electron:	1	2	3	4	5	6
	$n = 2$	$n = 2$	$n = 2$	$n = 2$	$n = 2$	$n = 2$
	$l = 1$	$l = 1$	$l = 1$	$l = 1$	$l = 1$	$l = 1$
	$m = +1$	$m = 0$	$m = -1$	$m = +1$	$m = 0$	$m = -1$
	$s = +\frac{1}{2}$	$s = +\frac{1}{2}$	$s = +\frac{1}{2}$	$s = -\frac{1}{2}$	$s = -\frac{1}{2}$	$s = -\frac{1}{2}$.

These six sets of quantum numbers represent six electrons residing in the $2p$ level, which consists of three orbitals, each holding two electrons. Each value of m denotes an orbital that can hold two electrons with $s = +\frac{1}{2}$ and $s = -\frac{1}{2}$. This was the case for the $1s$ and $2s$ orbitals, but in those cases m was restricted to the value 0 because $l = 0$ for an s orbital.

TABLE 5.1 ▶ Maximum Numbers of Electrons That States Can Hold

l value	m values	State	Maximum number of electrons
0	0	s	2
1	0, ± 1	p	6
2	0, ± 1, ± 2	d	10
3	0, ± 1, ± 2, ± 3	f	14
4	0, ± 1, ± 2, ± 3, ± 4	g	18

For a given value of l, there are always as many orbitals as there are m values, with each orbital capable of holding a pair of electrons. Thus, for $l = 3$ there are seven possible values for m (0, ± 1, ± 2, and ± 3) so that such an f state can hold 14 electrons. Table 5.1 shows the number of electrons that the various states can hold.

For convenience, we always write sets of quantum numbers by using the highest positive value of m first and working down. We will also start with the positive value of s first. Thus, for Al, the "last" electron is in the $3p$ level and we will assign it the set of quantum numbers $n = 3$, $l = 1$, $m = 1$, and $s = +\frac{1}{2}$.

Except for minor variations, the order of increasing energy levels in an atom is given by the sum $(n + l)$. The lowest value for $(n + l)$ occurs when $n = 1$ and $l = 0$ for the $1s$ state. The next lowest sum of $(n + l)$ is 2, which occurs when $n = 2$ and $l = 0$. We cannot have a $1p$ state where $n = 1$ and $l = 1$ because of the restrictions on n and l that arise from the solution of the wave equation. For $(n + l) = 3$, the possible combinations are $n = 2$ and $l = 1$ $(2p)$ and $n = 3$ and $l = 0$ $(3s)$. Although the sum $(n + l)$ is the same in both cases, the level with $n = 2$ is filled first. Therefore, we conclude that when two or more ways exist for the same sum $(n + l)$ to arise, the level with the lower n will usually fill first. Thus, the approximate order of filling the energy states in atoms is shown in Table 5.2.

We can describe the filling of energy states and the maximum occupancies of the orbitals by making use of the order shown earlier. The filling of the orbitals is regular until Cr is reached. There, we predict $3d^4\ 4s^2$ but find $3d^5\ 4s^1$. The reason for this is the more favorable coupling of spin and orbital angular momenta that results for $3d^5\ 4s^1$, which has six unpaired electrons. Coupling of angular momenta will be discussed in the next section.

The relationship of electronic structure to the periodic table should be readily apparent. Groups IA and IIA represent the groups where an s level is being filled as the outer shell. The first, second, and third series of transition

TABLE 5.2 ▶ Energy States According to the *(n + l)* Sum

n	l	$(n + l)$	State	
1	0	1	$1s$	
2	0	2	$2s$	
2	1	3	$2p$	Increasing energy ↓
3	0	3	$3s$	
3	1	4	$3p$	
4	0	4	$4s$	
3	2	5	$3d$	
4	1	5	$4p$	
5	0	5	$5s$	
4	2	6	$4d$	
5	1	6	$5p$	
6	0	6	$6s$	
4	3	7	$4f$	
5	2	7	$5d$	
6	1	7	$6p$	
7	0	7	$7s$	

elements are the groups where the $3d$, $4d$, and $5d$ levels are being filled. As a result, such elements are frequently referred to as d group elements. The main group elements to the right in the periodic table represent the periods where $2p$, $3p$, $4p$, $5p$, and $6p$ levels are the outside shells in the various long periods. Finally, the rare earths and the actinides represent groups of elements where the $4f$ and $5f$ levels are being filled.

The periodic table shows the similarities in electron configurations of elements in the same group. For example, the alkali metals (Group IA) all have an outside electron arrangement (valence shell) of ns^1, where $n = 2$ for Li, $n = 3$ for Na, etc. Since the chemical properties of elements depend on the outer-shell electrons, it is apparent why elements in this group are similar chemically. By adding one electron, the halogens (Group VIIA), having configurations $ns^2\ np^5$, are converted to the configuration of the next noble gas, $ns^2\ np^6$. We should emphasize, however, that although there are many similarities, numerous differences also exist between elements in the same group. Thus, it should not be inferred that the same electron configuration in the valence shell gives rise to the same chemical properties for all members of the group.

5.4 Spectroscopic States

After the overall electronic configuration of an atom has been determined, there are still other factors that affect the energy state. For example, the electronic configuration of carbon is $1s^2\ 2s^2\ 2p^2$. The fact that two electrons are indicated in the $2p$ state is insufficient for a complete description of the atom since there are several ways in which those electrons may be arranged. This description does not take into account either electron repulsion or spin–orbit coupling. In other words, the electronic configuration alone is based only on the n and l quantum numbers. Therefore, within the $1s^2\ 2s^2\ 2p^2$ configuration there are several different energy states.

The energy states arise because the orbital and spin angular momentum vectors may couple to provide several states of different energy. Two ways in which vector coupling can occur will be discussed. These represent limiting cases, and intermediate coupling schemes are known. In the first coupling scheme, the individual orbital moments (l) couple to give a resultant orbital angular moment, L. Also, the individual spin moments (s) couple to produce a total spin moment, S. The two vector quantities, L and S, then couple to give the total angular momentum quantum number, J. This scheme is known as Russell–Saunders or L–S coupling. In this case, the coupling of individual spin moments and individual orbital moments is stronger than the coupling between individual spin and orbital moments. In the other extreme, the individual spin and orbital moments for a given electron couple to produce a resultant, j, for that electron. These j vectors then couple to produce the resultant, J, the overall angular momentum. Coupling of this type is called j–j coupling.

For relatively light atoms, L–S coupling provides the better model. The j–j coupling scheme occurs for heavier atoms in the lower part of the periodic table. For our purposes, the L–S coupling scheme will suffice. We determine the coupling of spin and orbital momenta according to a scheme that we will now describe.

An electron in an atom is characterized by a set of four quantum numbers. The orbital angular momentum quantum number, l, gives the length of the orbital angular momentum vector in units of $h/2\pi$. The actual quantum mechanical result is $[l(l+1)]^{1/2}$ instead of l. This is because the quantity $l(l+1)$ is the square of the eigenvalue of the operator for the z component of angular momentum. Although the correct value is $[l(l+1)]^{1/2}$, l is commonly used since a vector of length $[l(l+1)]^{1/2}$ has exactly the same quantized projections on the z axis as a vector of l units in length.

Because the overall angular momentum is produced by the coupling of vectors, we need to see how the coupling can occur. Figure 5.2 shows the coupling of two vectors, $l = 1$ and $l = 2$, according to quantization. It is readily apparent that for two vectors of lengths $l = 1$ and $l = 2$, the resultant, L, can be formed according to quantum restrictions in three ways. Note that the values of L are $|l_1 + l_2|$, $|l_1 + l_2 - 1|$, and $|l_1 - l_2|$. We will have occasion to use this quantum mechanical coupling of vectors in arriving at the overall angular momentum state.

As we mentioned earlier, the quantum number m gives the projection of the orbital angular momentum vector on the z axis. This vector can precess around the z axis, sweeping out cones of revolution around that axis. This is shown in Figure 5.3 for $l = 2$ (for which $[l(l+1)]^{1/2} = (6)^{1/2}$), which has projections on the z axis of 0, ±1, and ±2. In the absence of an external magnetic field, these orientations are degenerate.

If spin–orbit coupling occurs, it is not possible to say what the value of the orbital component will be merely from the orbital angular momentum quantum number. This number can give only the maximum value of the projection on the z axis. Consequently, the *microstates* (detailed arrangements of the electrons) must be written in order to predict the spin–orbit

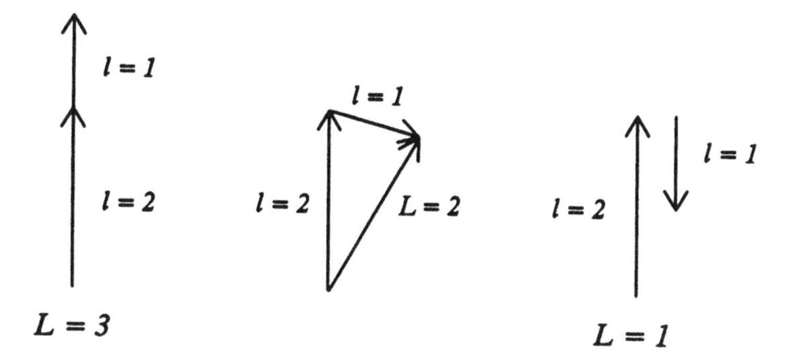

Figure 5.2 ▶ Quantized combinations of vectors $l = 2$ and $l = 1$.

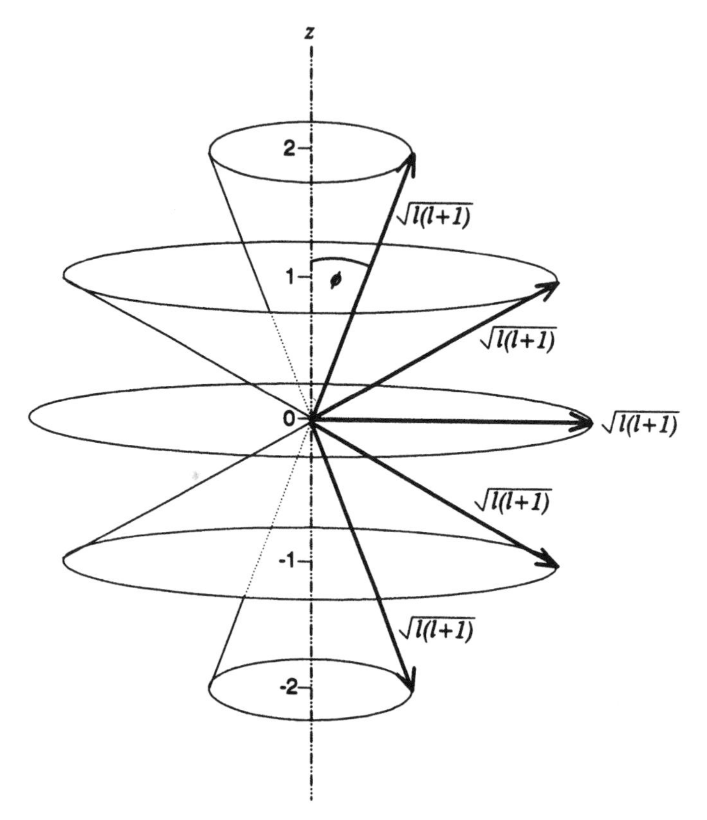

Figure 5.3 ▶ The projections on the z axis of a vector that is $[l(l+1)]^{1/2}$ units in length (the case for $l = 2$). The angular momentum vector can precess around the z axis and has projections of $+2$, $+1$, 0, -1, and -2.

coupling. However, this is not necessary if only the lowest energy state in desired.

The essential aspect of spin–orbit coupling is that the spin angular momenta for two or more unpaired electrons couple to form a resultant spin vector, S. For the configuration

$$\underline{\uparrow} \quad \underline{\uparrow} \quad \underline{\quad}$$

$S = (\frac{1}{2}) + (\frac{1}{2}) = 1$. The orbital angular momenta couple similarly in their z projections. It is the sum of the m values that gives the maximum length of the L vector. These two resultant vectors, L and S, then couple to give a third vector, J, which can have all integral values from $|L + S|$ to $|L - S|$

(see Figure 5.2). We can summarize the rules as follows:

$$L = \sum l_i$$
$$S = \sum s_i$$
$$M = \sum m_i = L,\ L-1,\ L-2, \ldots, 0, \ldots, -L$$
$$J = |L+S|, \ldots, |L-S|.$$

Once the sums M are obtained, it should be obvious what L projections the M values represent. The L values 0, 1, 2, and 3 respectively correspond to the *states* designated as S, P, D, and F, in accord with the practice for electronic energy levels (s, p, d, f).

The *multiplicity* of a state is given by $(2S+1)$, where S is the sum of the spins. Thus, a single unpaired electron (as in Na, $3s^1$) gives rise to a *doublet* $[2(\frac{1}{2})+1=2]$. A *term symbol* is written as $(2S+1)L_J$, with some examples being $^2S_0$, $^2P_2$, and $^3D_1$. There are also three rules that will enable us to determine the relative energy of the terms once they are obtained. These rules are known as *Hund's rules*:

1. For equivalent electrons, the state with the highest multiplicity gives the lowest energy

2. Of those states having the highest multiplicity, the one with the highest L is lowest in energy

3. For shells less than half filled, the lowest J gives the state of lowest energy; for shells more than half-filled, the highest J gives the lowest energy.

A few examples will be used to illustrate how the rules work.

Consider first the case of a configuration ns^1. In this case, $S=\frac{1}{2}$ and $L=0$ because the m value for an s state is 0. Therefore, the state is designated as ^2S in accord with the rules. Only a single J value is possible, that being $\frac{1}{2}$. Therefore, the spectroscopic state (term) is $^2S_{1/2}$. This is the only term possible for a configuration of ns^1.

Consider next the case of ns^2. In this case, $S=0$ since the two electrons have opposite spins and $L=0$ because the electrons reside in an orbital for which $l=0$ (hence $m=0$ and the sum of the $m_i=0$). The only possible value for J is 0, so that the ground state term is $^1S_0$. Actually, this must be the result for any filled shell, p^6, d^{10}, etc. Because filled shells contribute only 0 to the S and L values, we can ignore them when determining the spectroscopic state and consider only the partially filled outer shell.

Consider now the np^1 configuration. In this case, there are three orbitals with m values of $+1, 0$, and -1 where the electron may be found. These projections on the z axis can arise only for a vector $L = 1$ since $L = \Sigma m_i = 1$. The ground state must be a P state ($L = 1$). For a single unpaired electron, $S = \frac{1}{2}$, so the multiplicity is equal to $(2S + 1) = 2$. For $L = 1$ and $S = \frac{1}{2}$, two values are possible for the vector J. These are $\left|1 + (\frac{1}{2})\right| = \frac{3}{2}$ and $\left|1 - (\frac{1}{2}) = \frac{1}{2}\right|$. Therefore, the two spectroscopic states that exist for the np^1 configuration are $^2P_{1/2}$ and $^2P_{3/2}$, with $^2P_{1/2}$ being lower in energy since the p shell is less than half filled. For the boron atom, the $^2P_{1/2}$ state lies 16 cm^{-1} lower than the $^2P_{3/2}$ state, while for the aluminum atom the difference is 112 cm^{-1}.

Let us now examine the case of the np^2 configuration. We shall begin by writing the 15 microstates shown in Table 5.3 that are possible for this configuration. It is apparent that the highest value of L is 2. This occurs with $S = 0$, and hence the vector $L = 2$ can have five projections on the z axis. These will be given by the series of M values of $2, 1, 0, -1$, and -2, the states designated in the table by **. These 5 states constitute a ^1D term. Of the remaining microstates, the highest M value is 1 and the highest S is 1. Therefore, these must represent a ^3P state. Actually, since we have all possible combinations of $M = 1, 0$, and -1 with $S = 1, 0$, and -1, there are a total of 9 microstates used to make the ^3P term. The corresponding J values are given by $|1 + 1|$, $|1 + 1 - 1|$, and $|1 - 1|$, so that the values

TABLE 5.3 ▶ The Microstates Arising from the np^2 Configuration

$m = +1$	$m = 0$	$m = -1$	$S = \Sigma s_i$	$L = \Sigma m_i$	Label
↑	↑		1	1	*
↑		↑	1	0	*
	↑	↑	1	−1	*
↓	↓		−1	1	*
↓		↓	−1	0	*
	↓	↓	−1	−1	*
↓	↑		0	1	*
↓		↑	0	0	*
	↓	↑	0	−1	*
↑↓			0	2	**
↑	↓		0	1	**
	↑↓		0	0	**
	↑	↓	0	−1	**
		↑↓	0	−2	**
↑		↓	0	0	***

TABLE 5.4 ▶ Relative Energy Levels (in cm^{-1}) for Terms Arising from the np^2 Configuration

	Atom				
State	Carbon	Silicon	Germanium	Tin	Lead
$^3P_0$	0.0	0.0	0.0	0.0	0.0
$^3P_1$	16.4	77.2	557.1	1,691.8	7,819.4
$^3P_2$	43.5	223.3	1,409.9	3,427.7	10,650.5
$^1D_2$	10,193.7	6,298.8	7,125.3	8,613.0	21,457.9
$^1S_0$	21,648.4	15,394.2	16,367.1	17,162.6	29,466.8

are 2, 1, and 0. Thus, we have used the 9 microstates designated by * in the table. One microstate indicated by *** in the table, which has $M = 0$ and $S = 0$, remains. This combination can only correspond to the term $^1S_0$.

We have found the terms associated with the np^2 configuration to be $^3P_0$, $^3P_1$, $^3P_2$, $^1D_2$, and $^1S_0$, with Hund's rules predicting exactly that order of increasing energy. The energy levels relative to the $^3P_0$ ground state are shown in Table 5.4.

In the lighter elements carbon and silicon, the coupling scheme is of the *L–S* type. For elements as heavy as lead, the *j–j* coupling scheme is followed. Note that while all of the elements have a ground state of $^3P_0$, the singlet terms are much higher in energy.

In many instances we are interested in obtaining only the ground state term. This can be done without going through the complete procedure just outlined for the np^2 configuration. For example, Hund's rules indicate that the state with the highest multiplicity will be lowest in energy. Consequently, to determine the ground state term we need only to look at the states where the sum of spins is highest, which results when the electrons are unpaired and have the same spin. For the np^2 configuration, the states are those where $S = 1$ and $L = 1$. This corresponds to a ^3P state. We then work out the J values as before. This amounts to simply placing electrons in orbitals with the highest m values and working down while placing the electrons in the orbitals with the same spin. Therefore, for np^2,

$$m = \underset{+1}{\underline{\uparrow}} \quad \underset{0}{\underline{\uparrow}} \quad \underset{-1}{\underline{\quad}}.$$

Immediately we see that $S = 1$ and $L = 1$, which leads us directly to a multiplicity of 3, a P state, and J values of 2, 1, and 0. Thus, we have found quickly that the ground state for the np^2 configuration is $^3P_0$.

For the d^1 configuration, $S = \frac{1}{2}$ and the maximum M value is 2 because $l = 2$ for a d state. Thus, the ground state term will be a ^2D term with a J

value of $\frac{3}{2}$. There are other terms corresponding to higher energies, but in this way the ground state term is found easily. Consider a d^2 configuration. We could consider the arrangements

$$m = \frac{\uparrow}{+2} \quad \frac{\uparrow}{+1} \quad \frac{}{0} \quad \frac{}{-1} \quad \frac{}{-2}$$

or

$$m = \frac{\uparrow\downarrow}{+2} \quad \frac{}{+1} \quad \frac{}{0} \quad \frac{}{-1} \quad \frac{}{-2}.$$

Hund's rules tell us that the first of these will lie lower in energy. Consequently, the *maximum* value of Σm_i is 3 when the electrons are unpaired, so the L vector is 3 units long ($L = 3$ corresponds to an F state). Therefore, the term lying lowest in energy is ^3F with a J value of 2 or $^3F_2$.

One further point should be mentioned. If a set of degenerate orbitals can hold x electrons, the terms arising from $(x - y)$ electrons in those orbitals are exactly the same as the terms arising for y electrons in those orbitals. That is to say that permuting a vacancy among the orbitals produces the same effect as permuting an electron among the orbitals. Thus, a p^4 configuration gives rise to the same term as a p^2 configuration. Only the order of J values is inverted for the shell that is greater than half filled.

TABLE 5.5 ► Spectroscopic Terms for Equivalent Electrons[a]

Electron configuration	Terms
s^1	^2S
s^2	^1S
p^1	^2P
p^2	$^3P, {}^1D, {}^1S$
p^3	$^4S, {}^2D, {}^2P$
p^4	$^3P, {}^1D, {}^1S$
p^5	^2P
p^6	^1S
d^1	^2D
d^2	$^3F, {}^3P, {}^1G, {}^1D, {}^1S$
d^3	^4F (ground state only)
d^4	^5D (ground state only)
d^5	^6S (ground state only)
d^6	^5D (ground state only)
d^7	^4F (ground state only)
d^8	^3F (ground state only)
d^9	^2D
d^{10}	^1S

[a] The J values have been omitted for simplicity.

References for Further Reading

- Alberty, R. A., and Silbey, R. J. (1996). *Physical Chemistry*, 2nd ed. Wiley, New York. Solid coverage of angular momenta and coupling.
- Eyring, H., Walter, J., and Kimball, G. E. (1944). *Quantum Chemistry*, Chap. 9. Wiley, New York. A theoretical approach to the structure of complex atoms.
- Hameka, H. F. (1967). *Introduction to Quantum Theory*, Chaps. 10 and 11. Harper & Row, New York. A clear presentation of the principles of perturbation theory and the variation method.
- Herzberg, G. (1944). *Atomic Spectra and Atomic Structure*. Dover, New York. The best known source of information on line spectra and coupling in complex atoms.
- Laidler, K. J., and Meiser, J. H. (1982). *Physical Chemistry*, Chap. 11. Benjamin–Cummings, Menlo Park, CA. A good readable account of atomic structure.
- Pauling, L. (1960). *The Nature of the Chemical Bond*, Chap. 2. Cornell Univ. Press, Ithaca, NY. A wealth of information on atomic structure.
- Pauling, L., and Wilson, E. B. (1935) *Introduction to Quantum Mechanics*. McGraw–Hill, New York. Theoretical development of perturbation theory and variation method presented in considerable detail.

Problems

1. Determine by means of vector diagrams the possible values that the total angular momentum can have for the following combinations:

   (a) $L = 3$ and $S = \frac{5}{2}$; and

   (b) $L = 2$ and $S = \frac{5}{2}$.

2. Use a vector diagram to show what total angular momenta that a single electron in an f state can have.

3. For each of the following, determine the ground state spectroscopic term and sketch the splitting pattern that would result for the atom in a magnetic field:

   (a) Al,

   (b) P,

   (c) Ca,

   (d) Ti, and

   (e) Se.

4. Write a set of four quantum numbers for the "last" electron in each of the following:

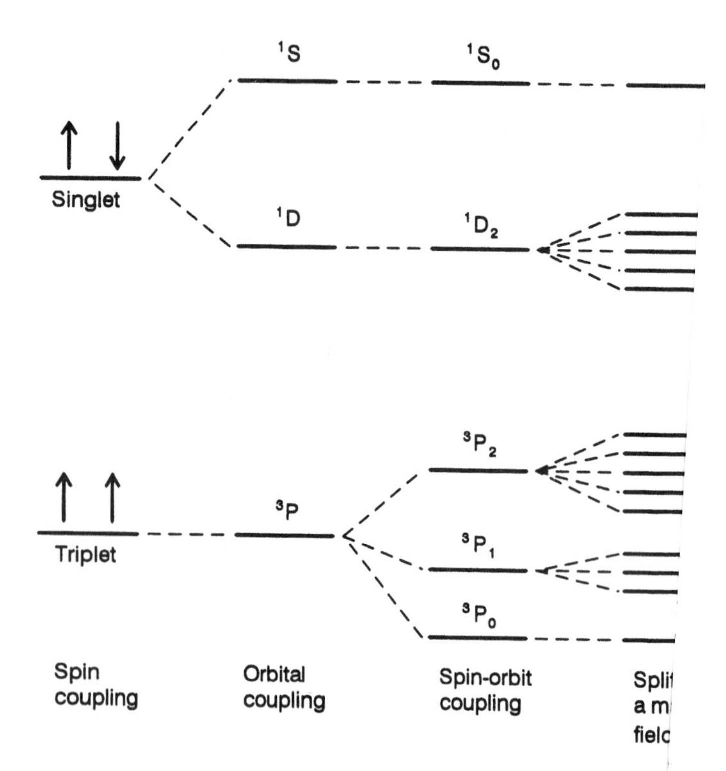

Figure 5.4 ▶ Splitting of states due to spin–orbit coupling. There are
magnetic field corresponding to projections of the J vector in the z dire

Table 5.5 shows the terms arising from the various electror
The significance of the energies represented by the variou
shown in Figure 5.4. The inclusion of electron repulsion s
state into the various triplet and singlet states arising fror
tion. Further, the inclusion of spin–orbit interactions sepa
into as many components as there are J values. Finally
of a magnetic field removes the degeneracy within the ^{3}
to produce multiplets represented by the different orienta
the J vector. Figure 5.4 shows a summary of these state
spin–orbit coupling and the effect of an external magneti

While we have not been able to solve Schrödinger w
complex atoms, this chapter shows the approach taken fo
We have also described electron configurations and the v
atom that leads to coupling of angular momenta. This ir
be supplemented by consultation of the references liste
coverage is needed.

(**a**) Sc,

(**b**) Cl,

(**c**) Sr,

(**d**) V, and

(**e**) Co.

5. Calculate the most probable radius of He^+ using $P(r) = 4\pi r [R(r)]^2 r^2$.

6. The ionization potential for the potassium atom is 4.341 eV. Estimate the effective atomic number, Z_{eff}, for the 4s electron in potassium.

7. Using Slater's rules, determine the effective nuclear charge for the fifth electron in the 2p level of fluorine. Write the wave function for this electron.

8. Determine the most probable distance for the electron in the 1s state of Be^{3+}.

9. Determine the spectroscopic term for the ground state for the following:

(**a**) Ti^{3+},

(**b**) Cr^{3+},

(**c**) O^{2-},

(**d**) Mn^{2+}, and

(**e**) Ni^{2+}.

10. Calculate $\langle r \rangle$ for the electron in the 1s state of Li^{2+}. Use the wave function shown in Table 4.1.

▶ **Chapter 6**

Vibrations and the Harmonic Oscillator

Most of what we know about the structure of atoms and molecules has been obtained by studying the interaction of electromagnetic radiation with matter or its emission from matter. The vibrations in molecular systems constitute one of the properties that provides a basis for studying molecular structure by spectroscopic techniques. *Infrared* spectroscopy provides the experimental technique for studying changes in vibrational states in molecules, and that technique is familiar to chemistry students even at a rather low level. Molecular vibrations in gaseous molecules also involve changes in rotational states so changes in these types of energy levels are sometimes considered together. In this chapter, we will develop some of the concepts required for an interpretation of molecular vibrations and their study using spectroscopic experiments.

We will begin with a discussion of an object vibrating on a spring to show some of the physical concepts and mathematical techniques. Solving problems related to vibrations requires the use of differential equations, and some persons studying quantum mechanics for the first time may not have taken such a course or may need a review. Consequently, this chapter also includes a very limited coverage of this area of mathematics that is so important in the physical sciences.

6.1 The Vibrating Object

We will first deal with a simple problem in vibrational mechanics by considering the arrangement shown in Figure 6.1.

For an object attached to a spring, *Hooke's law* describes the system in terms of the force on the object and the displacement from the equilibrium position:

$$F = -kx. \tag{6.1}$$

In this equation, x is the distance the object is displaced from its equilibrium position, and k is known as the *spring constant* or *force constant*, which is expressed in dimensions of force/distance. Appropriate units for k are dynes/centimeter, Newtons/meter, or, in the case of chemical bonds, millidynes/ängstrom. The negative sign indicates that the restoring force (spring tension) is in the direction opposite to the displacement. The energy (or work) required to cause the displacement (which is the potential energy given the object) is expressed by the force law integrated over the interval that the spring is stretched:

$$\int_0^x F(x)\,dx = \int_0^x -kx\,dx = \frac{1}{2}kx^2. \tag{6.2}$$

If the mass m is displaced by a distance of x and released, the object vibrates in simple harmonic motion. The *angular frequency* of the vibration, ω, will be given by

$$\omega = \sqrt{\frac{k}{m}}, \tag{6.3}$$

while the *classical* or *vibrational frequency*, ν, is given by

$$\nu = \frac{1}{2\pi}\sqrt{\frac{k}{m}}. \tag{6.4}$$

From Eqs. (6.3) and (6.4) it is clear that $\omega = 2\pi\nu$.

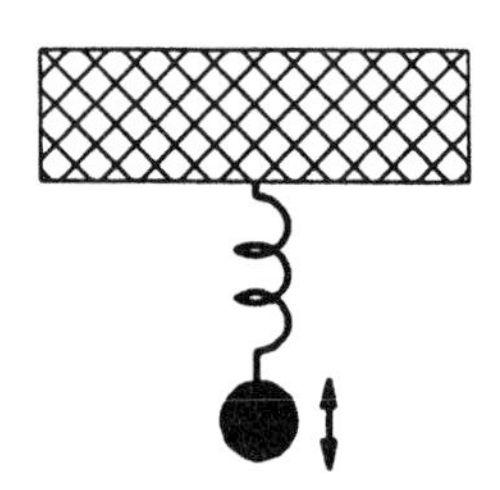

Figure 6.1 ▶ An object vibrating on a spring.

The maximum displacement from the equilibrium position is called the *amplitude*, and the variation of the displacement with time is found by making use of Newton's Second Law of Motion, $F = ma$. Velocity is the first derivative of distance with time, dx/dt, and acceleration is the derivative of velocity with time, d^2x/dt^2. Therefore, we can write $F = ma$ as

$$m\frac{d^2x}{dt^2} = -kx \tag{6.5}$$

or

$$\frac{d^2x}{dt^2} + \frac{k}{m}x = 0, \tag{6.6}$$

which is a linear differential equation with constant coefficients. Before progressing to the solution of this problem in vibrations, we will present a brief discussion of the type of differential equation involved in the analysis.

6.2 Linear Differential Equations with Constant Coefficients

In this Section, we present the results of several important theorems in differential equations. Because of the nature of this book, we present them in an operational manner without proof. The interested reader should consult a text on differential equations for more details.

A linear differential equation with constant coefficients is an equation of the form

$$a_n(x)\frac{d^ny}{dx^n} + a_{n-1}(x)\frac{d^{n-1}y}{dx^{n-1}} + \cdots + a_1(x)\frac{dy}{dx} + a_0(x)y = F(x), \tag{6.7}$$

where the constants $a_0(x), a_1(x), \ldots,$ and $F(x)$ have values that change only with x. A particularly important equation of this type is the second-order case,

$$a_2(x)\frac{d^2y}{dx^2} + a_1(x)\frac{dy}{dx} + a_0(x)y = F(x). \tag{6.8}$$

The *differential operator*, D, is defined as

$$D = \frac{d}{dx},\ D^2 = \frac{d^2}{dx^2},\ \text{etc.} \tag{6.9}$$

When an operator meets the conditions that

$$D(f + g) = Df + Dg \tag{6.10}$$

and

$$D^n(f + g) = D^n f + D^n g, \tag{6.11}$$

the operator is called a *linear operator*.

A second-order linear differential equation can be written in the operator notation as

$$a_2 D^2 y + a_1 D y + a_0 = F(x). \tag{6.12}$$

The solution of an equation of this form is obtained by considering an *auxiliary equation*, which is obtained by writing an equation in the form

$$f(D)y = 0 \tag{6.13}$$

when the general differential equation is written as

$$f(D)y = F(x). \tag{6.14}$$

The auxiliary equation is called the *complementary equation* and its solution is known as the *complementary solution*. The *general solution* of the differential equation is the sum of the particular solution of the general equation and the solution of the complementary equation. We will illustrate all this by the following example: Suppose we wish to find the general solution of

$$\frac{d^2y}{dx^2} - 5\frac{dy}{dx} + 4y = 10x. \tag{6.15}$$

In operator form this equation becomes

$$(D^2 - 5D + 4)y = 10x. \tag{6.16}$$

A solution of this type of equation is frequently of the form

$$y = C_1 e^{ax} + C_2 e^{bx}, \tag{6.17}$$

with a and b being determined by the solutions of the complementary equation,

$$m^2 - 5m + 4 = 0. \tag{6.18}$$

Therefore, factoring the polynomial gives

$$(m - 4)(m - 1) = 0, \tag{6.19}$$

from which we find $m = 4$ and $m = 1$. In this case, the general solution of Eq. (6.15) is

$$y = C_1 e^{x} + C_2 e^{4x}. \tag{6.20}$$

We can easily verify the solution by using it in the complementary equation. If this is the solution, then

$$Dy = \frac{dy}{dx} = C_1 e^x + 4C_2 e^{4x}$$
$$D^2 y = \frac{d^2 y}{dx^2} = C_1 e^x + 16C_2 e^{4x}.$$

Now we can write the auxiliary equation as

$$(D^2 - 5D + 4)y = D^2 y - 5Dy + 4y = 0. \tag{6.21}$$

By substitution we obtain

$$C_1 e^x + 16C_2 e^{4x} - 5(C_1 e^x + 4C_2 e^{4x}) + 4(C_1 e^x + C_2 e^{4x}) = 0, \tag{6.22}$$

which reduces to $0 = 0$. However, it can also be shown that a *particular* solution is

$$y = \frac{5}{2}x + \frac{25}{8}, \tag{6.23}$$

and it also satisfies the general equation. In this case,

$$Dy = \frac{dy}{dx} = \frac{5}{2} \quad \text{and} \quad D^2 y = \frac{d^2 y}{dx^2} = 0, \tag{6.24}$$

so substituting these values in Eq. (6.15) gives

$$-5\left(\frac{5}{2}\right) + 4\left(\frac{5x}{2} + \frac{25}{8}\right) = 10x$$
$$10x = 10x.$$

Therefore, the complete solution of Eq. (6.15) is the sum of the two expressions,

$$y = C_1 e^x + C_2 e^{4x} + \frac{5}{2}x + \frac{25}{8}. \tag{6.25}$$

In most problems, we are content with a general solution and "singular" solutions that do not describe the physical behavior of the system are ignored.

It must be mentioned that there are two arbitrary constants that characterize the solution that we obtained. Of course, an nth-order equation results in n constants. In quantum mechanics, these constants are determined by the physical constraints of the system (known as boundary conditions), as we saw in Chapter 3.

The equation

$$D^2 y + y = 0 \tag{6.26}$$

has the auxiliary equation

$$m^2 + 1 = 0 \tag{6.27}$$

so that $m^2 = -1$ and $m = \pm i$. The general solution is

$$y = C_1 e^{ix} + C_2 e^{-ix}. \tag{6.28}$$

At this point, it is useful to remember that

$$\frac{d}{dx}(\sin x) = \cos x \tag{6.29}$$

and

$$\frac{d}{dx}(\cos x) = -\sin x = \frac{d^2}{dx^2}(\sin x). \tag{6.30}$$

Therefore,

$$\frac{d^2}{dx^2}(\sin x) + \sin x = 0, \tag{6.31}$$

and the solution $y = \sin x$ satisfies the equation. In fact, if we assume a solution to Eq. (6.26) of the form

$$y = A \sin x + B \cos x, \tag{6.32}$$

then

$$Dy = A \cos x - B \sin x \tag{6.33}$$

$$D^2 y = -A \sin x - B \cos x. \tag{6.34}$$

Therefore,

$$D^2 y + y = -A \sin x - B \cos x + A \sin x + B \cos x = 0 \tag{6.35}$$

and the solution shown in Eq. (6.32) satisfies Eq. (6.26). A differential equation can have only one general solution so the solution in Eqs. (6.28) and (6.32) must be equal

$$y = C_1 e^{ix} + C_2 e^{-ix} = A \sin x + B \cos x. \tag{6.36}$$

When $x = 0$, $C_1 + C_2 = B$. Differentiating Eq. (6.36),

$$\frac{dy}{dx} = C_1 i e^{ix} - C_2 i e^{-ix} = A \cos x - B \sin x. \tag{6.37}$$

Now at $x = 0$, $\sin x = 0$ and it is apparent that

$$i(C_1 - C_2) = A. \tag{6.38}$$

Substituting for A and B and simplifying, we obtain

$$C_1 e^{ix} + C_2 e^{-ix} = C_1(\cos x + i \sin x) + C_2(\cos x - i \sin x). \tag{6.39}$$

If $C_2 = 0$ and $C_1 = 1$,

$$e^{ix} = \cos x + i \sin x, \tag{6.40}$$

and if $C_2 = 1$ and $C_1 = 0$,

$$e^{-ix} = \cos x - i \sin x. \tag{6.41}$$

The relationships shown in Eqs. (6.40) and (6.41) are known as *Euler's formulas*.

Suppose we wish to solve the equation

$$y'' + 2y' + 5y = 0, \tag{6.42}$$

which can be written in operator form as

$$(D^2 + 2D + 5)y = 0. \tag{6.43}$$

The auxiliary equation is

$$m^2 + 2m + 5 = 0, \tag{6.44}$$

and its roots are found by using the quadratic formula:

$$m = \frac{-2 \pm \sqrt{4 - 20}}{2} = -1 \pm 2i. \tag{6.45}$$

Therefore, the solution of Eq. (6.42) is

$$y = C_1 e^{(-1+2i)x} + C_2 e^{(-1-2i)x} \tag{6.46}$$

or

$$y = C_1 e^{-x} e^{2ix} + C_2 e^{-x} e^{-2ix} = e^{-x}(C_1 e^{2ix} + C_2 e^{-2ix}). \tag{6.47}$$

Using Euler's formulas we obtain

$$y = e^{-x}(A \sin 2x + B \cos 2x). \tag{6.48}$$

In general, if the auxiliary equation has roots $a \pm bi$, the solution of the differential equation has the form

$$y = e^{ax}(A \sin bx + B \cos bx). \tag{6.49}$$

An equation of the form

$$y'' + a^2 y = 0 \tag{6.50}$$

arises in solving the particle in the one-dimensional box problem in quantum mechanics (see Chapter 3). The auxiliary equation is

$$m^2 + a^2 = 0, \tag{6.51}$$

which has the solutions $m = \pm ia$. Therefore, the solution of Eq. (6.50) can be written as

$$y = C_1 e^{aix} + C_2 e^{-aix} = A \cos ax + B \sin ax, \tag{6.52}$$

which is exactly the form of the solution found earlier. The boundary conditions make it possible to evaluate the constants in the solution as was shown in Chapter 3.

6.3 Back to the Vibrating Object

Let us now return to the vibrating system described in Section 6.1. Suppose a force of 6.0 N stretches the spring 0.375 m. We will take the displacement to be negative so that the spring constant, k, is

$$k = -\frac{f}{x} = -\frac{6.0 \text{ N}}{-0.375 \text{ m}} = 16 \text{ N/m}. \tag{6.53}$$

Now let us assume that the object has a mass of 4.00 kg and it is raised 0.375 m above its equilibrium position as shown in Figure 6.2 and released.

As shown earlier, the motion of the object is described by the equation

$$\frac{d^2x}{dt^2} + \frac{k}{m}x = 0 \tag{6.54}$$

or, making use of the data given in this case and the condition $x(0) = 0.375$,

$$\frac{d^2x}{dt^2} + \frac{16}{4}x = 0. \tag{6.55}$$

$$\frac{d^2x}{dt^2} + \frac{k}{m}x = 0$$

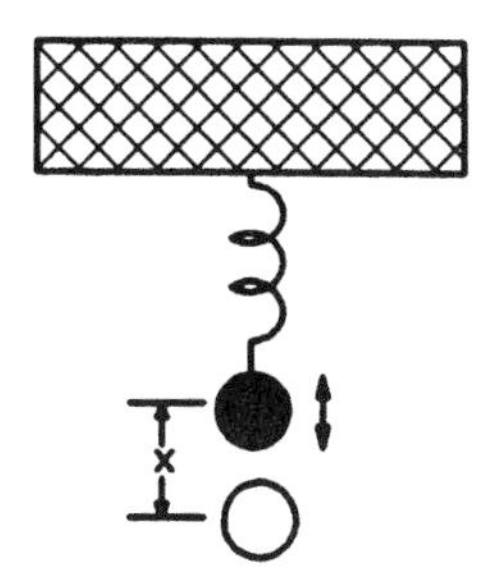

Figure 6.2 ▶ An object vibrating after displacing it from its normal position.

The auxiliary equation is

$$m^2 + 4 = 0, \tag{6.56}$$

so $m^2 = -4$ and $m = \pm 2i$. The general solution to Eq. (6.54) can be written as

$$x = C_1 e^{2it} + C_2 e^{-2it} = A \sin 2t + B \cos 2t. \tag{6.57}$$

At the beginning of the motion, $t = 0$ and $x = 0.375$ m, and the velocity of the object is 0 or $v = dx/dt = 0$. Therefore,

$$x = A \sin 2t + B \cos 2t = 0.375 \text{ m}, \tag{6.58}$$

but since $\sin 0 = 0$ and $\cos 0 = 1$,

$$0.375 = B \cos 0 = B \cdot 1. \tag{6.59}$$

Consequently, $B = 0.375$ and the partial solution is

$$x = 0.375 \cos 2t + A \sin 2t. \tag{6.60}$$

Taking the derivative, dx/dt,

$$\frac{dx}{dt} = -0.375(2) \sin 2t + 2A \cos 2t = 0. \tag{6.61}$$

When $t = 0$, the $\sin \theta$ term goes to 0 but

$$2A \cos 2t = 2A \cos 0 = 2A \cdot 1 = 0. \tag{6.62}$$

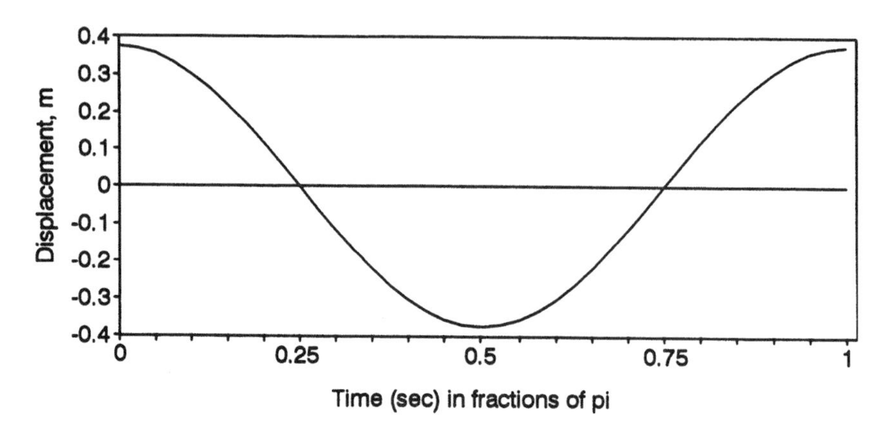

Figure 6.3 ▶ The variation of displacement with time for a complete cycle.

Therefore, A must be equal to 0. The required solution is

$$x = 0.375 \cos 2t. \tag{6.63}$$

Figure 6.3 shows the graphical nature of this solution with the displacement of the object being 0.375 m at $t = 0$ and varying as a cosine function thereafter. The period of the vibration is the time necessary for one complete vibration while the frequency is given by

$$\nu = \frac{1}{2\pi}\sqrt{\frac{k}{m}} = \frac{1}{2\pi}\sqrt{\frac{16(\text{kg m/s}^2)/\text{m}}{4\text{ kg}}} = \frac{1}{\pi}\text{ s}^{-1}.$$

The problem of the classical vibrating object serves to introduce the terminology and techniques used for the quantum mechanical oscillator. The latter is a much more complex problem, which we will now consider.

6.4 The Quantum Mechanical Harmonic Oscillator

One of the very useful models in quantum mechanics is the harmonic oscillator. This model provides the basis for discussing vibrating chemical bonds (discussed further in Chapter 7) so it is a necessary part of the discussion of spectroscopy. We will now describe this important quantum mechanical model.

Earlier, it was shown that for a vibrating object the potential energy is given by

$$V = \frac{1}{2}kx^2, \tag{6.64}$$

but it can also be written as

$$V = \frac{1}{2}mx^2\omega^2, \tag{6.65}$$

where ω is the angular frequency of vibration, $(k/m)^{1/2}$. The classical vibrational frequency, ν, is $(1/2\pi)(k/m)^{1/2}$. From these definitions, we can see that $\omega = 2\pi\nu$.

The total energy of the oscillator is the sum of the potential and kinetic energies. In order to write the Schrödinger equation, we need to find the form of the Hamiltonian operator. The kinetic energy, T, can be written in operator form as

$$\hat{T} = -\frac{\hbar^2}{2m}\frac{d^2}{dx^2}. \tag{6.66}$$

The potential energy is $2\pi^2\nu^2mx^2$, so the Hamiltonian operator can be written as

$$\hat{H} = -\frac{\hbar^2}{2m}\frac{d^2}{dx^2} + 2\pi^2\nu^2mx^2. \tag{6.67}$$

If $b = 2\pi\nu m/\hbar$, then $\hat{H}\psi = E\psi$ and this equation can be written as

$$\frac{d^2\psi}{dx^2} + \left(\frac{2mE}{\hbar^2} - b^2x^2\right)\psi = 0. \tag{6.68}$$

In this equation, the potential varies with x^2, and since it is a nonlinear function, this equation is much more complex than that of the classical harmonic oscillator or the particle in the one-dimensional box.

Inspection of the wave equation shows that the solution must be a function such that its second derivative contains both the original function *and* a factor of x^2. A function like $\exp(-bx^2)$ satisfies that requirement. In fact, we will show later that the solution can be written as

$$\psi = c[\exp(-bx^2)], \tag{6.69}$$

where b and c are constants, and that it is possible to show that this solution satisfies Eq. (6.68). The solution of the equation by rigorous means requires that it be solved by a method using infinite series. Before applying this technique to an equation of the complexity of Eq. (6.68), we will illustrate its use in simple cases.

6.5 Series Solutions of Differential Equations

A large number of problems in science and engineering are formulated in terms of differential equations that have variable coefficients. The quantum mechanical harmonic oscillator is one such problem, but we also saw such equations in the solution of the hydrogen atom problem in Chapter 4. At that time, the solutions were simply stated without any indication of how we solve such equations. Some of the most famous equations of this type are shown in Table 6.1. In these equations, n or v represents a constant.

The equations shown in Table 6.1 are differential equations for which the solutions are given as infinite series or polynomial solutions, usually bearing the name of the person who solved the problem. We certainly do not intend to solve all of these equations, but they are given to show some of the "name" differential equations that will be encountered in a more advanced study of quantum mechanics. They constitute some of the most important differential equations to be encountered in theoretical work. It should be mentioned that these equations can be written in other forms, so they may not be readily recognized. We have already seen the commonly used technique in quantum mechanics of manipulating an equation to get it in a recognizable form in Chapter 4.

Since the solution of the harmonic oscillator problem will be given in somewhat greater detail than was given in treating the hydrogen atom problem, we will begin by illustrating the solution of differential equations by means of series. This is done for the reader whose mathematics background includes calculus but not differential equations. For a more complete discussion of this technique, see the references at the end of the chapter. Some of the differential equations that we have already solved have had solutions

TABLE 6.1 ► Some Important Nonlinear Differential Equations That Are Solved by a Series Technique

Name	Equation	Solutions
Hermite	$y'' - 2xy' + 2ny = 0$	Hermite polynomials
Bessel	$y'' + \frac{1}{x}y' + \left(1 - \frac{v^2}{x^2}\right)y = 0$	Bessel functions
Legendre	$y'' - \frac{2x}{1-x^2}y' + \frac{v(v+1)}{1-x^2}y = 0$	Legendre polynomials
Chebyshev	$y'' - \frac{x}{1-x^2}y' + \frac{n^2}{1-x^2}y = 0$	Chebyshev polynomials
Laguerre	$x^2y'' + (1-x)y' + ny = 0$	Laguerre polynomials

that were trigonometric or exponential functions. Since these functions can be written as series, the solutions of such equations could have been written in the form of series. Consider the equation

$$\frac{dy}{dx} = y, \tag{6.70}$$

with $y = 1$ at $x = 0$. This equation can be solved by inspection because the only function we know that equals its first derivative is e^x. Also, we could write the equation as

$$\frac{dy}{dx} - y = 0, \tag{6.71}$$

which has the auxiliary equation

$$m - 1 = 0 \tag{6.72}$$

so that $m = 1$, and using the techniques illustrated in Section 6.2, we can write the solution as e^x. Suppose the previous approaches were not taken but rather that a solution was *assumed* so that

$$y = a_0 + a_1x + a_2x^2 + a_3x^3 + a_4x^4 + \cdots. \tag{6.73}$$

Substitution of this series in Eq. (6.70) gives

$$\frac{d}{dx}(a_0 + a_1x + a_2x^2 + a_3x^3 + \cdots) = \tag{6.74}$$
$$a_0 + a_1x + a_2x^2 + a_3x^3 + \cdots,$$

$$a_1 + 2a_2x + 3a_3x^2 + \cdots = a_0 + a_1x + a_2x^2 + a_3x^3 + \cdots, \tag{6.75}$$

which is true if coefficients of like powers of x are equal. So,

$$a_0 = a_1 \qquad a_1 = 2a_2 \qquad a_2 = 3a_3 \qquad a_3 = 4a_4$$
$$a_2 = \frac{a_1}{2} = \frac{a_0}{2!} \qquad a_3 = \frac{a_2}{3} = \frac{a_0}{6} = \frac{a_0}{3!} \qquad a_4 = \frac{a_3}{4} = \frac{a_0}{4!}.$$

Therefore, since the assumed solution is

$$y = a_0 + a_1x + a_2x^2 + a_3x^3 + a_4x^4 \cdots, \tag{6.76}$$

substituting the coefficients found earlier and factoring out a_0 gives

$$y = a_0\left(1 + x + \frac{x^2}{2!} + \frac{x^3}{3!} + \frac{x^4}{4!} + \cdots\right). \tag{6.77}$$

From the initial condition that $y = 1$ when $x = 0$ we see that $y = a_0 = 1$. Therefore, $a_0 = 1$ and the required solution is

$$y = 1 + x + \frac{x^2}{2!} + \frac{x^3}{3!} + \frac{x^4}{4!} + \cdots . \qquad (6.78)$$

It is interesting to note that

$$e^x = 1 + x + \frac{x^2}{2!} + \frac{x^3}{3!} + \frac{x^4}{4!} + \cdots , \qquad (6.79)$$

which is in agreement with the solution given earlier. Of course it is not always this simple, but it is reassuring to see that the series solution is exactly the same as that already known from other methods.

To provide another, less obvious illustration of the method, let us consider the equation

$$\frac{dy}{dx} = xy, \quad \text{with } y(0) = 2. \qquad (6.80)$$

Assuming as before that

$$y = a_0 + a_1 x + a_2 x^2 + a_3 x^3 + \cdots , \qquad (6.81)$$

we find that the derivative dy/dx is

$$\frac{dy}{dx} = a_1 + 2a_2 x + 3a_3 x^2 + 4a_4 x^3 + 5a_5 x^4 + 6a_6 x^5 + 7a_7 x^6 + \cdots . \qquad (6.82)$$

The product xy becomes

$$xy = a_0 x + a_1 x^2 + a_2 x^3 + a_3 x^4 + a_4 x^5 + a_5 x^6 + a_6 x^7 + \cdots . \qquad (6.83)$$

Therefore, from the original equation, Eq. (6.80), we see that

$$\begin{aligned} a_1 + 2a_2 x + 3a_3 x^2 + 4a_4 x^3 + \cdots \\ = a_0 x + a_1 x^2 + a_2 x^3 + a_3 x^4 + a_4 x^5 + \cdots . \end{aligned} \qquad (6.84)$$

Equating coefficients of like powers of x, we find that

$$\begin{array}{lllll} 2a_2 = a_0 & 3a_3 = a_1 & 4a_4 = a_2 & 5a_5 = a_3 & 6a_6 = a_4 \\ a_2 = \frac{a_0}{2} & a_3 = 0 & a_4 = \frac{a_2}{4} = \frac{a_0}{2 \cdot 4} & a_5 = 0 & a_6 = \frac{a_4}{6} = \frac{a_0}{2 \cdot 4 \cdot 6} = \frac{a_0}{2^3 \cdot 3!} . \end{array}$$

From the condition that $y(0) = 2$, we see that $a_0 = 2$ and that a_1 must be 0 because there is no term with a corresponding power of x in Eq. (6.84).

Therefore, $a_3 = a_5 = 0$ and substituting the preceding values for the coefficients gives

$$y = 2 + 0x + \frac{a_0}{2}x^2 + 0x^3 + \frac{a_0}{2 \cdot 4}x^4 + 0x^5 + \frac{a_0}{2 \cdot 4 \cdot 6}x^6 + \cdots . \quad (6.85)$$

Because $a_0 = 2$, substituting for a_0 and factoring out 2 gives

$$y = 2\left(1 + \frac{x^2}{2!} + \frac{x^4}{2^3} + \frac{x^6}{2^3 \cdot 3!} + \cdots\right) = 2\left[\exp\left(\frac{x^2}{2}\right)\right]. \quad (6.86)$$

The preceding discussion is intended to show how series solutions are obtained for relatively simple differential equations. The equations that arise in the quantum mechanical treatment of problems are more complex than these, but the brief introduction is sufficient to remove some of the mystery of using series in this way. It is not expected that the reader will know how to solve the complex equations of mathematical physics, but the approach will now be familiar even if the details are not.

6.6 Back to the Harmonic Oscillator

In the previous sections, we have examined simple problems dealing with vibrations and the differential equations that describe them. As we shall see, the quantum mechanical harmonic oscillator is a quite different problem. As we saw earlier, we can formulate the problem in a relatively simple fashion (see Figure 6.4). Using Hooke's law, the restoring force is written as

$$F = -kx \quad (6.87)$$

and the potential energy, V, is given by

$$V = \frac{1}{2}kx^2. \quad (6.88)$$

The angular frequency of vibration is $\omega = (k/m)^{1/2}$, so $k = m\omega^2$ and

$$V = \frac{1}{2}m\omega^2 x^2. \quad (6.89)$$

We know that the total energy, E, is the sum of the kinetic and potential energies,

$$E = T + V. \quad (6.90)$$

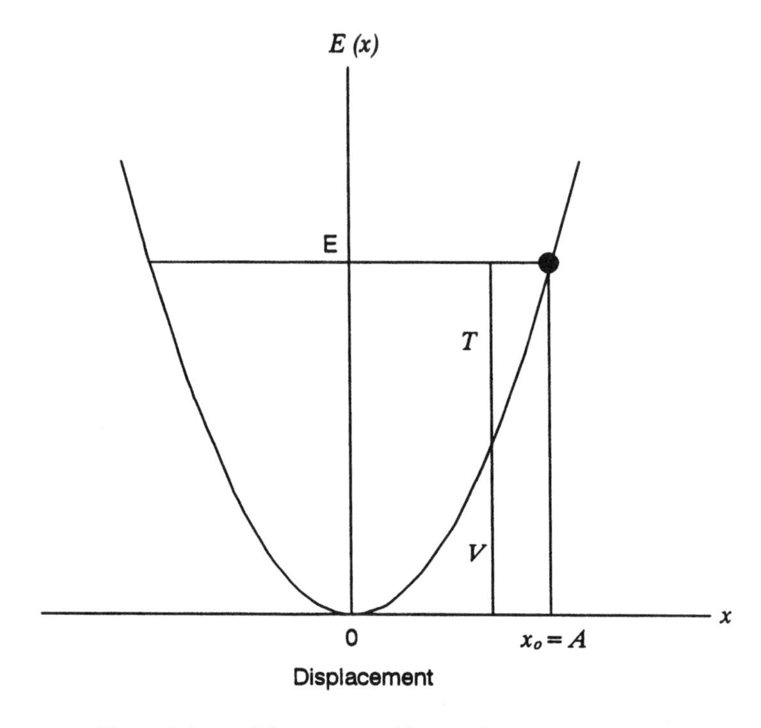

Figure 6.4 ▶ A harmonic oscillator following Hooke's law.

At the equilibrium position of the vibration, $V = 0$ and T is a maximum, and the total energy is $E = T$. At the extremes of the vibration, the oscillator comes to rest before it changes direction and the total energy is the potential energy, $E = V$. It is near the extremes of the vibration where the velocity is low that the oscillator spends the majority of its time. Near the equilibrium position, where the velocity is the highest, the probability of finding the oscillator is lowest. As we will discuss later, the probability of finding the oscillator at any point along its motion is inversely proportional to the velocity at that point. The total energy can be written in terms of the amplitude, A, as

$$E = \frac{1}{2}kA^2 = \frac{1}{2}m\omega^2 A^2. \tag{6.91}$$

The potential energy can be expressed as

$$V = \frac{1}{2}m\omega^2 x^2. \tag{6.92}$$

As usual, we write the Schrödinger equation as

$$\hat{H}\psi = E\psi, \tag{6.93}$$

and the Hamiltonian operator in this case is

$$\hat{H} = -\frac{\hbar^2}{2m}\frac{\partial^2}{\partial x^2} + V = -\frac{\hbar^2}{2m}\frac{\partial^2}{\partial x^2} + \frac{1}{2}m\omega^2 x^2. \tag{6.94}$$

Therefore, $\hat{H}\psi = E\psi$ becomes

$$\left(-\frac{\hbar^2}{2m}\frac{\partial^2}{\partial x^2} + \frac{1}{2}m\omega^2 x^2\right)\psi = E\psi. \tag{6.95}$$

Simplifying this equation by multiplying by $-2m$ and dividing by $\hbar^2$ gives

$$\left(\frac{\partial^2}{\partial x^2} - \frac{m^2\omega^2 x^2}{\hbar^2}\right)\psi = -\frac{2mE}{\hbar^2}\psi, \tag{6.96}$$

which can be written as

$$\frac{\partial^2\psi}{\partial x^2} = -\left(\frac{2mE}{\hbar^2} - \frac{m^2\omega^2}{\hbar^2}x^2\right)\psi. \tag{6.97}$$

This is *not* a linear differential equation, and we will have greater difficulty obtaining a solution. If we assume for the moment that the solution has the form

$$\psi = c\left[\exp\left(-bx^2\right)\right], \tag{6.98}$$

where b and c are constants, we can check this solution in Eq. (6.97). We begin by taking the required derivatives,

$$\frac{d\psi}{dx} = -2bxc\left[\exp\left(-bx^2\right)\right] \tag{6.99}$$

$$\frac{d^2\psi}{dx^2} = -2bc\left[\exp\left(-bx^2\right)\right] + 4b^2cx^2\left[\exp\left(-bx^2\right)\right]. \tag{6.100}$$

Now, working with the right-hand side of Eq. (6.97), we find that

$$-\left(\frac{2mE}{\hbar^2} - \frac{m^2\omega^2}{\hbar^2}x^2\right)\psi = -\frac{2mE}{\hbar^2}c\left[\exp\left(-bx^2\right)\right] + \frac{m^2\omega^2}{\hbar^2}x^2c\left[\exp\left(-bx^2\right)\right]. \tag{6.101}$$

We note that both Eqs. (6.100) and (6.101) contain terms in x^2 and terms that do not contain x except in the exponential. Therefore, we can equate terms that contain x^2,

$$\frac{m^2\omega^2}{\hbar^2}x^2c\left[\exp\left(-bx^2\right)\right] = 4b^2cx^2\left[\exp\left(-bx^2\right)\right]. \tag{6.102}$$

Canceling common factors from both sides, we obtain

$$4b^2 = \frac{m^2\omega^2}{\hbar^2} \qquad (6.103)$$

or

$$b = \frac{m\omega}{2\hbar}. \qquad (6.104)$$

Working with the terms that do not contain x as a factor, we find that

$$E = b\frac{\hbar^2}{m}, \qquad (6.105)$$

and substituting for b we find that

$$E = \frac{1}{2}\omega\hbar. \qquad (6.106)$$

Therefore, when $b = m\omega/2\hbar$ and $E = \omega\hbar/2$, the function

$$\psi = c\left[\exp\left(-bx^2\right)\right] \qquad (6.107)$$

satisfies the Schrödinger equation for a harmonic oscillator. Using the value obtained for b, we can write the solution as

$$\psi = c\left[\exp\left(\frac{-m\omega x^2}{2\hbar}\right)\right]. \qquad (6.108)$$

This is, in fact, the solution for the harmonic oscillator in its lowest energy state. Although we have assumed that the solution has this form, we now need to show how the problem is solved.

The solution of the harmonic oscillator problem will now be addressed, starting with the wave equation written as

$$\frac{d^2\psi}{dx^2} + \frac{2m}{\hbar^2}\left(E - \frac{1}{2}kx^2\right)\psi = 0. \qquad (6.109)$$

If we let $\alpha = 2mE/\hbar^2$ and $\beta = 2\pi(mk)^{1/2}/h$, Eq. (6.109) becomes

$$\frac{d^2\psi}{dx^2} + \left(\alpha - \beta^2x^2\right)\psi = 0. \qquad (6.110)$$

The usual trick now employed is to introduce a change in variable such that

$$z = \sqrt{\beta}x; \qquad (6.111)$$

then the second derivatives are related by

$$\frac{d^2}{dx^2} = \beta \frac{d^2}{dz^2}. \tag{6.112}$$

The wave equation can now be written as

$$\beta \frac{d^2\psi}{dz^2} + \left(\alpha - \beta^2 x^2\right)\psi = 0 \tag{6.113}$$

or

$$\frac{d^2\psi}{dz^2} + \left(\frac{\alpha}{\beta} - \beta x^2\right)\psi = 0. \tag{6.114}$$

Therefore, since $z^2 = \beta x^2$,

$$\frac{d^2\psi}{dz^2} + \left(\frac{\alpha}{\beta} - z^2\right)\psi = 0. \tag{6.115}$$

If we express the solution as a function of z,

$$\psi\,(z) = u\,(z)\exp\left(-\frac{z^2}{2}\right), \tag{6.116}$$

using the simplified notation that $\psi = \psi\,(z)$ and $u = u\,(z)$, we can obtain the necessary derivatives as

$$\psi' = u' \exp\left(-\frac{z^2}{2}\right) - uz\exp\left(-\frac{z^2}{2}\right)$$

and

$$\begin{aligned}\psi'' &= u'' \exp\left(-\frac{z^2}{2}\right) - u'z\exp\left(-\frac{z^2}{2}\right) - u'z\exp\left(-\frac{z^2}{2}\right) \\ &\quad - u\exp\left(-\frac{z^2}{2}\right) + uz^2\exp\left(-\frac{z^2}{2}\right).\end{aligned} \tag{6.117}$$

Simplifying gives

$$\begin{aligned}\psi'' &= u'' \exp\left(-\frac{z^2}{2}\right) - 2u'z\exp\left(-\frac{z^2}{2}\right) \\ &\quad - u\exp\left(-\frac{z^2}{2}\right) + uz^2\exp\left(-\frac{z^2}{2}\right).\end{aligned} \tag{6.118}$$

Making the substitutions of Eqs. (6.116) and (6.118) in Eq. (6.115), we obtain

$$\frac{d^2u}{dz^2} - 2z\frac{du}{dz} + \left(\frac{\alpha}{\beta} - 1\right)u = 0. \tag{6.119}$$

If we let the factor $[(\alpha/\beta) - 1]$ be $2n$, Eq. (6.119) becomes

$$\frac{d^2u}{dz^2} - 2z\frac{du}{dz} + 2nu = 0, \tag{6.120}$$

which has exactly the form of Hermite's equation (Table 6.1).

Before considering the solution of Eq. (6.120) by means of a series, we will consider the energy levels for the harmonic oscillator. Since

$$\frac{\alpha}{\beta} - 1 = 2n, \tag{6.121}$$

then

$$\frac{\alpha}{\beta} = 2n + 1. \tag{6.122}$$

However, we know that

$$\frac{\alpha}{\beta} = \frac{8\pi mE}{2h\sqrt{mk}} = \frac{4\pi E\sqrt{m}}{h\sqrt{k}}. \tag{6.123}$$

Therefore,

$$2n + 1 = \frac{4\pi E}{h}\sqrt{\frac{m}{k}}. \tag{6.124}$$

Solving for the energy, we obtain

$$E = \frac{h}{2\pi}\sqrt{\frac{k}{m}}\left(n + \frac{1}{2}\right) = \hbar\sqrt{\frac{k}{m}}\left(n + \frac{1}{2}\right). \tag{6.125}$$

Since the frequency of vibration is $\nu = (1/2\pi)(k/m)^{1/2}$ and $\omega = 2\pi\nu$, we can write the expression for the energy as

$$E = \hbar\omega\left(n + \frac{1}{2}\right). \tag{6.126}$$

This equation is analogous to Eq. (6.106), which applied only to the ground state. The quantized energy levels arise from the restrictions on the nature of Hermite's equation.

Although we have given the preceding integer (quantum number) as n, the *vibrational* quantum number is usually designated as V. Therefore, the vibrational energy levels of the harmonic oscillator can be expressed in terms of the quantum number, V,

$$E = (V + \frac{1}{2})\hbar\omega \qquad (V = 0, 1, 2, \ldots), \tag{6.127}$$

and this results in a series of energy levels as shown in Figure 6.5. The spacing between the energy levels is $\hbar\omega$ and there is a zero point energy at $(1/2)\hbar\omega$. In 1900 Planck's treatment of blackbody radiation (see Chapter 1) predicted the same arrangement of energy levels. Almost 30 years later the quantum mechanical solution of the harmonic oscillator problem gave the same result.

The solution of Hermite's equation by a polynomial series will now be briefly addressed. Because of the nature of this equation and the complexity of its solution, we will provide an outline of the methods used. An advanced

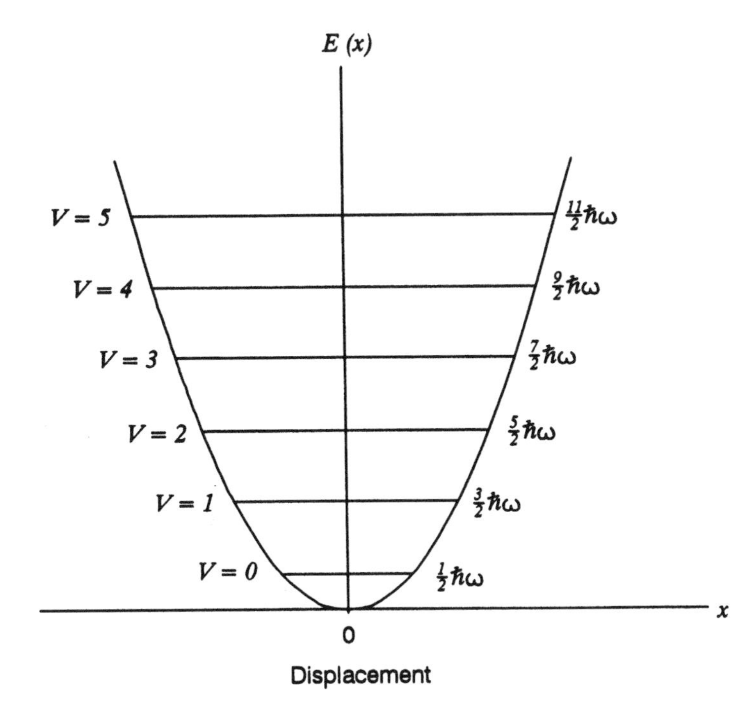

Figure 6.5 ▶ The quantized energy levels of the harmonic oscillator.

book on differential equations should be consulted for details of the solution of this type of equation. The Hermite equation is written as

$$\frac{d^2u}{dz^2} - 2z\frac{du}{dz} - 2pz = 0, \tag{6.128}$$

where p is an integer. We will assume a series solution as

$$H(z) = a_0 + a_1 z + a_2 z^2 + \cdots = \sum_{p=0}^{\infty} a_p z^p. \tag{6.129}$$

The required derivatives are

$$H'(z) = a_1 + 2a_2 z + 3a_3 z^2 + \cdots = \sum_{p=0}^{\infty} p a_p z^{p-1} \tag{6.130}$$

$$H''(z) = 2a_2 + 6a_3 z + 12a_4 z^2 + \cdots = \sum_{p=0}^{\infty} p(p-1) a_p z^{p-2}. \tag{6.131}$$

The terms involving a_0 and a_1 do not occur in the summation for $H''(z)$. Therefore, the series can be written as

$$H''(z) = \sum_{p=0}^{\infty} (p+1)(p+2) a_{p+2} z^p. \tag{6.132}$$

Using these results, the Hermite equation can now be written as

$$\sum_{p=0}^{\infty} \left[(p+1)(p+2) a_{p+2} + (2n - 2p) a_p\right] z^p = 0. \tag{6.133}$$

For this equation to be true for all values of z, the function in brackets must be zero:

$$\left[(p+1)(p+2) a_{p+2} + (2n - 2p) a_p\right] = 0. \tag{6.134}$$

Solving for a_{p+2},

$$a_{p+2} = \frac{-(2n-2p)}{(p+1)(p+2)} a_p. \tag{6.135}$$

This is the *recursion formula* for the coefficients of the series to be calculated. Using this formula,

$$\text{for } p = 0, \quad a_{p+2} = a_2 = -\frac{2\,(n-0)}{(1)\,(2)}a_0 = -na_0$$

$$\text{for } p = 1, \quad a_{p+2} = a_3 = -\frac{n-1}{3}a_1$$

$$\text{for } p = 2, \quad a_{p+2} = a_4 = -\frac{[2n-2\,(2)]}{(3)\,(4)}a_2 = -\frac{n-2}{6}a_2$$

$$= \frac{n\,(n-2)}{6} = a_0$$

$$\text{for } p = 3, \quad a_{p+2} = a_5 = -\frac{n-3}{10}a_0 = \frac{(n-1)\,(n-3)}{30}a_1.$$

Two constants, a_0 and a_1, are not given by the recursion relation. As we saw earlier, these are the two arbitrary constants that result from the solution of a second-order differential equation.

The next step in the solution is to show that the series can be written in terms of $\exp(-z^2)$ and that appropriate values can be assigned to the constants a_0 and a_1 to result in a well-behaved wave function. We will not go through that rather tedious process since it is best left to more advanced texts. However, the Hermite polynomials can be written as

$$H_n(z) = (-1)^n \exp\left(z^2\right) \frac{d^n}{dz^n} \exp\left(-z^2\right). \tag{6.136}$$

The first few Hermite polynomials can be written as

$$\begin{aligned} H_0\,(z) &= 1 \\ H_1\,(z) &= 2z \\ H_2\,(z) &= 4z^2 - 2 \\ H_3\,(z) &= 8z^3 - 12z \\ H_4\,(z) &= 16z^4 - 48z^2 + 12. \end{aligned} \tag{6.137}$$

The wave functions for the harmonic oscillator, ψ_i are written as a normalization constant, N_i, times $H_i(z)$ to give

$$\begin{aligned} \psi_0 &= N_0 \exp\left(-z^2\right) \\ \psi_1 &= N_1\,(2z) \exp\left(-z^2\right) \\ \psi_2 &= N_2\left(4z^2 - 2\right) \exp\left(-z^2\right) \\ \psi_3 &= N_3\left(8z^3 - 12z\right) \exp\left(-z^2\right). \end{aligned} \tag{6.138}$$

We have only outlined the mathematical procedure necessary to obtain the full solution of the harmonic oscillator model using quantum mechanics. The details are not really required for a discussion at this level, but the general approach should be appreciated from the brief introduction to the solution of differential equations by series.

The first three wave functions and their squares are shown in Figure 6.6. The squares of the wave functions, which are proportional to the probability density, show that the oscillator is not restricted to the classical limits of the

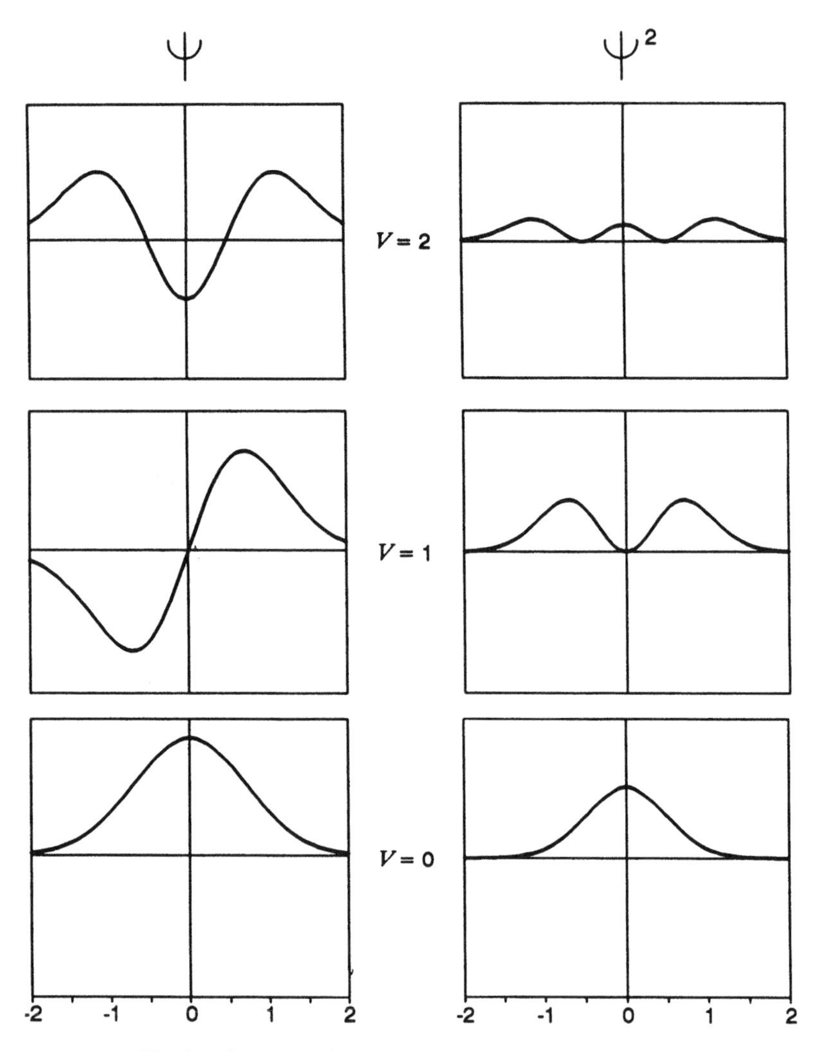

Figure 6.6 ▶ The first three wave functions and their squares for the harmonic oscillator.

vibration. For example, the plot of ψ^2 for the $V = 0$ state shows that there is a slight but finite probability that the oscillator can be found beyond the classical limit of ± 1, the range of the vibration. For states with $V > 0$, the probability of the oscillator tunneling through the classical limit is even greater (see Chapter 8).

It is interesting to note that as the position of a classical oscillator changes, its velocity continuously changes and reaches zero at the extremes of the vibration. The velocity has a maximum value at the center of the vibration (equilibrium position). Therefore, the time spent by the oscillator (and the probability of finding it) varies with position and has a minimum at the equilibrium distance.

We can analyze the relationship between probability and displacement for a classical oscillator in the following way: Reference to Figure 6.4 shows that the potential and kinetic energies vary during the oscillation while the total energy is constant. The total energy is

$$E = \frac{1}{2}kx_0^2, \tag{6.139}$$

while the potential energy is

$$V = \frac{1}{2}kx^2. \tag{6.140}$$

The kinetic energy, T, is given as

$$T = E - V = \frac{1}{2}kx_0^2 - \frac{1}{2}kx^2 = \frac{1}{2}k(x_0^2 - x^2) = \frac{1}{2}mv^2. \tag{6.141}$$

The velocity, v, is found from Eq. (6.141),

$$v = \sqrt{\frac{k}{m}}(x_0^2 - x^2)^{1/2}. \tag{6.142}$$

The probability of finding the oscillator at a given point is inversely proportional to its velocity,

$$P \sim \frac{1}{v} \sim \frac{1}{\left(x_0^2 - x^2\right)^{1/2}}. \tag{6.143}$$

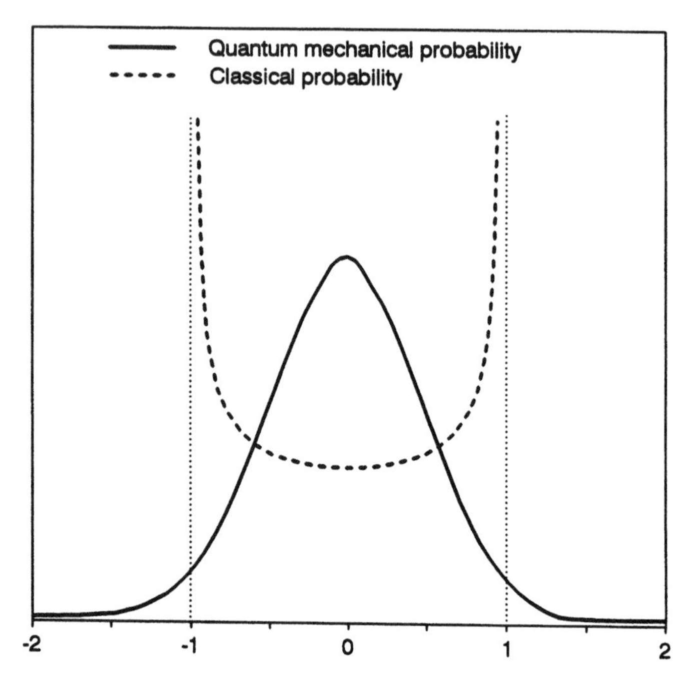

Figure 6.7 ▸ Probability distribution for the harmonic oscillator in its lowest energy state.

We write this equation as

$$P \sim \frac{1}{\left(1 - q^2\right)^{1/2}}, \tag{6.144}$$

where $q = x/x_0$. Using this function, we can generate a plot of the probability of finding the oscillator as a function of q. Figure 6.7 shows this classical probability of finding the oscillator. For the lowest energy state shown in Figure 6.6 the probability of finding the oscillator is given by the plot of ψ^2. This function is also shown in Figure 6.7.

It is immediately obvious that the probabilities obtained classically and quantum mechanically are greatly different. For the states of higher energy, the probabilities become more similar, in agreement with the principle that the quantum behavior approaches classical behavior under these conditions (referred to as the correspondence principle). According to the classical view of the harmonic oscillator, the probability of finding the oscillator is greatest near the limits of the vibration at ± 1, and there is no probability that the oscillator is beyond those limits. According to quantum mechanics, the maximum probability density occurs at the equilibrium position and there is a small but finite probability of finding the oscillator beyond the limits of the vibration.

6.7 Population of States

States of unequal energy are unequally populated. This principle is one of the most important ones in dealing with systems consisting of atoms and molecules. Our experience tells us that it is true as it applies to a liquid and its vapor, reactants and a transition state they form, or electrons populating orbitals in an atom. The principle, known as the *Boltzmann distribution law*, gives the relative populations of two states as

$$\frac{n_1}{n_0} = e^{-\Delta E/kT}, \tag{6.145}$$

where n is a population, ΔE is the difference in energy between the two states, k is Boltzmann's constant, and T is the temperature (K). When the energy is given on a molar basis, the equation becomes.

$$\frac{n_1}{n_0} = e^{-\Delta E/RT}. \tag{6.146}$$

Strictly, this equation applies to situations where the states are single. If either of the states consists of a set of levels, the degeneracies of the states, g, are included:

$$\frac{n_1}{n_0} = \frac{g_1}{g_0} e^{-\Delta E/kT}. \tag{6.147}$$

For a harmonic oscillator with energy states separated by 2000 cm^{-1} (4.0×10^{-13} erg), the population of the first excited state (n_1) relative to the ground state (n_0) at 300 K is

$$\frac{n_1}{n_0} = \exp\left[\frac{-4.0 \times 10^{-13}\ \text{erg}}{1.38 \times 10^{-16}\ \text{erg/molecule K} \times 300\ \text{K}}\right]$$
$$= e^{-9.66} = 6.4 \times 10^{-5}.$$

At 600 K, the relative population is only 7.0×10^{-3}. It is clear that at any reasonable temperature, a collection of a large number of oscillators will be found almost totally in the ground state. If the oscillators represent vibrating molecules, this has some significant implications for spectroscopic studies on the molecules, as will be discussed in Chapter 7.

The quantum mechanical harmonic oscillator is most important for its representation of vibrating molecules. Because of the relationship between molecular vibrations and rotations, this application of the harmonic oscillator will be deferred until we have described the model known as the rigid rotor.

References for Further Reading

- Adamson, A. W. (1966). *A Textbook of Physical Chemistry*, 3rd ed., Chap. 19, Academic Press, Orlando. A good treatment of vibration and molecular spectroscopy.
- Barrow, G. M. (1962). *Introduction to Molecular Spectroscopy*. McGraw–Hill, New York. A standard source for beginning a study of spectroscopic methods.
- Harris, D. C., and Bertolucci, M. D. (1989). *Symmetry and Spectroscopy*, Chap. 3. Dover, New York. An enormously readable and useful treatment of all aspects of spectroscopy.
- Laidlaw, W. G. (1970). *Introduction to Quantum Concepts in Spectroscopy*. McGraw–Hill, New York. A good treatment of the quantum mechanical models that are useful for interpreting spectroscopy.
- Laidler, K. J., and Meiser, J. H. (1982). *Physical Chemistry*, Chap. 13. Benjamin–Cummings, Menlo Park, CA. An outstanding introduction to most types of spectroscopy.
- Sonnessa, A. J. (1966). *Introduction to Molecular Spectroscopy*. Reinhold, New York. A very good introductory book on the subject of molecular spectroscopy.
- Wheatley, P. J. (1959). *The Determination of Molecular Structure*. Oxford Univ. Press, London. A classic introduction to experimental determination of molecular structure. Excellent discussion of spectroscopic methods.
- Wilson, E. B., Decius, J. C., and Cross, P. C. (1955). *Molecular Vibrations*. McGraw–Hill, New York. The standard reference on analysis of molecular vibrations. Now widely available in an inexpensive edition from Dover.

Problems

1. Calculate the zero-point vibrational energies for –O–H and –O–D bonds. If a reaction of these bonds involves breaking them, what does this suggest about the relative rates of the reactions of –O–H and –O–D bonds? What should be the ratio of the reaction rates?

2. The OH stretching vibration in gaseous CH_3OH is at 3687 cm^{-1}. Estimate the position of the O–D vibration in CH_3OD.

3. If $y = x^2 + 5x + 2e^x$, evaluate the following where $D = d/dx$:

(a) $(D^2 + 4D + 2)y$,

(b) $(D + 4)y$,

(c) $(2D^3 + 4D)y$.

4. If $y = \sin 3x + 4\cos 2x$, evaluate the following:

(a) $(D^2 + 3D + 3)y$,

(b) $(D + 3)y$,

(c) $(2D^2 + 3D)y$.

5. Use the auxiliary equation method to solve the following:

(a) $\dfrac{d^2y}{dx^2} + 4\dfrac{dy}{dx} - 5y = 0$,

(b) $(4D^2 - 36)y = 0$,

(c) $\dfrac{d^2y}{dx^2} - y = 0,\ y(0) = 2$, and $y'(0) = -3$,

(d) $(D^2 - 3D + 2)y = 0,\ y(0) = -1$, and $y'(0) = 0$.

6. Use the series approach to solve the following:

(a) $y' + y = 0$, with $y(0) = 1$,

(b) $\dfrac{dy}{dx} - xy = 0$, with $y(0) = 2$,

(c) $\dfrac{d^2y}{dx^2} + y = 0$.

7. Wave functions for which $\psi(x) = \psi(-x)$ are symmetric while those for which $\psi(x) = -\psi(-x)$ are antisymmetric. Determine whether the first four normalized wave functions for the harmonic oscillator are symmetric or antisymmetric.

8. Find the normalization constant for the wave function corresponding to the lowest energy state of a harmonic oscillator, $\psi = N_0 \exp\left(-bx^2\right)$, where b is a constant.

▶ Chapter 7

Molecular Rotation and Spectroscopy

Spectroscopy of molecules involves changes in the energy levels associated with vibration and rotation. In fact, much of what we know about the structure of atoms and molecules has been obtained by studying the interaction of electromagnetic radiation with matter. In this chapter, we introduce this important topic after considering the quantum mechanical problem of rotation and its combination with vibration.

7.1 Rotational Energies

To introduce the principles associated with rotation, we will first consider the case of an object of mass m rotating around a fixed center as shown in Figure 7.1. In this case, we consider the center of the rotation as being stationary, but this is only for convenience. The moment of inertia, I, in this case is

$$I = mr^2, \tag{7.1}$$

where r is the radius of rotation and m is the mass of the object. The angular velocity, ω, is given as the change in angle, ϕ, with time,

$$\omega = \frac{d\phi}{dt}. \tag{7.2}$$

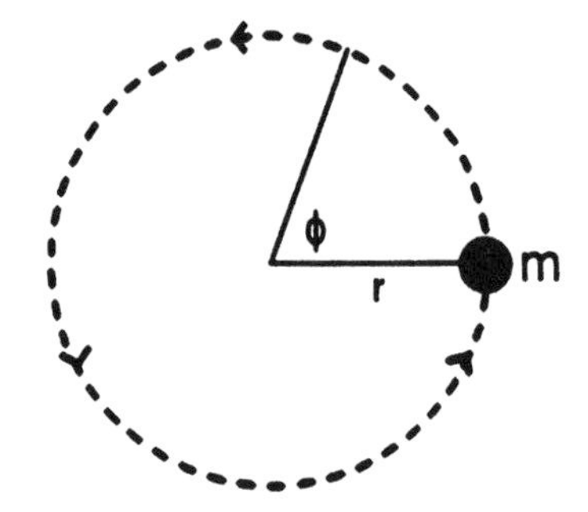

Figure 7.1 ▶ A rotating object of mass m.

The kinetic energy of rotation, T, is

$$T = \frac{1}{2} I \frac{d^2\phi}{dt^2}. \tag{7.3}$$

By analogy to linear momentum, the angular momentum can be written as

$$P_\phi = I \frac{d\phi}{dt} = I\omega, \tag{7.4}$$

and the kinetic energy is

$$T = \frac{P_\phi^2}{2I}, \tag{7.5}$$

which is analogous to $T = p^2/2m$ for linear momentum. In polar coordinates, the operator for angular momentum can be written as

$$\hat{P}_\phi = \frac{\hbar}{i} \frac{\partial}{\partial \phi}, \tag{7.6}$$

and the operator for rotational kinetic energy is

$$\hat{T} = -\frac{\hbar}{2I} \frac{\partial^2}{\partial \phi^2}. \tag{7.7}$$

Assuming that potential energy of the object is 0, $V = 0$, the Hamiltonian operator is $\hat{H} = \hat{T} + \hat{V} = \hat{T} + 0 = \hat{T}$ and the Schrödinger equation is

$$-\frac{\hbar}{2I} \frac{\partial^2 \psi}{\partial \phi^2} = E\psi, \tag{7.8}$$

which can be written as

$$\frac{\partial^2 \psi}{\partial \phi^2} + \frac{2I}{\hbar} E\psi = 0. \tag{7.9}$$

We have solved an equation of this form before in the particle in a one-dimensional box model. Therefore, the characteristic equation (where $k^2 = 2IE$) is

$$m^2 + k^2 = 0, \tag{7.10}$$

from which we find

$$m = \pm ik, \tag{7.11}$$

and we can write immediately

$$\psi = C_1 e^{ik\phi} + C_2 e^{-ik\phi}. \tag{7.12}$$

Since ϕ is an angular measure, $\psi(\phi) = \psi(\phi + 2\pi)$, where ϕ is in radians. Therefore,

$$(2IE)^{1/2} = J \quad (J = 0, 1, 2, \ldots) \tag{7.13}$$

or

$$E = \frac{J^2}{2I}. \tag{7.14}$$

If we follow the Bohr assumption that angular momentum is quantized, we can write $mvr = nh/2\pi$, where $I\omega = mvr$ is the angular momentum and n is an integer. Using J as the quantum number, we can write

$$I\omega = J\frac{h}{2\pi}. \tag{7.15}$$

Consequently,

$$\omega = \frac{hJ}{2\pi I}. \tag{7.16}$$

Substituting, we find that from Eq. (7.3),

$$E_{\text{rot}} = \frac{1}{2}I\omega^2 = \frac{1}{2}\frac{(I\omega)^2}{I}, \tag{7.17}$$

from which we obtain

$$E_{\text{rot}} = \frac{1}{2I}\left(\frac{hJ}{2\pi}\right)^2 = \frac{h^2}{8\pi^2 I}J^2. \tag{7.18}$$

This model is oversimplified in that it does not represent the rotation of a diatomic molecule around a center of mass as it is represented quantum mechanically and we *assumed* quantized angular momentum. In the next section, we will show how the rotation of a diatomic molecule is treated by quantum mechanics.

7.2 Quantum Mechanics of Rotation

In order to show the applicability of quantum mechanical methods to the rigid rotor problem, we will consider a diatomic molecule as shown in Figure 7.2. The bond in the molecule will be considered to be rigid so that molecular dimensions do not change. Because of the moments around the center of gravity,

$$m_1r_1 = m_2r_2, \tag{7.19}$$

where m is mass and r is the distance from the center of gravity. It is obvious that

$$R = r_1 + r_2. \tag{7.20}$$

By substitution, we find that

$$r_1 = \frac{m_2 R}{m_1 + m_2} \tag{7.21}$$

$$r_2 = \frac{m_1 R}{m_1 + m_2}. \tag{7.22}$$

The moment of inertia, I, is

$$I = m_1r_1^2 + m_2r_2^2. \tag{7.23}$$

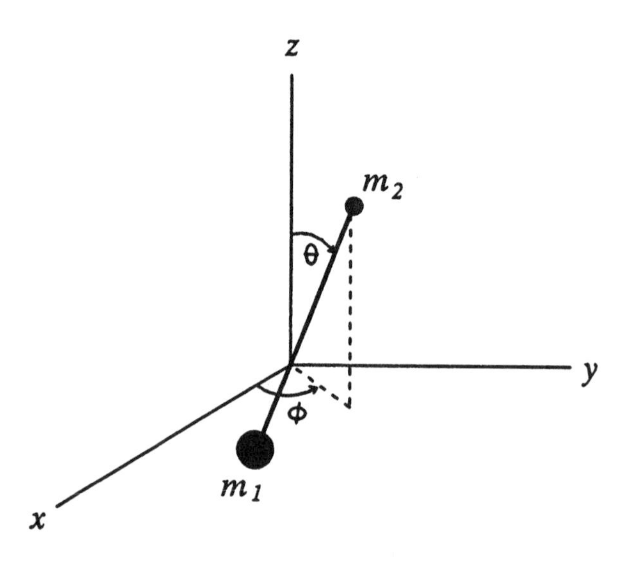

Figure 7.2 ▶ The rigid rotor coordinate system.

When the results obtained above for r_1 and r_2 are used, I can be written as

$$I = \mu R^2, \tag{7.24}$$

where μ is the reduced mass, $m_1 m_2/(m_1 + m_2)$.

For convenience, we will place the center of gravity at the origin of the coordinate system as shown in Figure 7.2. However, the solution of this problem is carried out by transforming the model into polar coordinates (see Chapter 4). For each atom, the kinetic energy can be written as

$$T = \frac{1}{2} m \left[\left(\frac{\partial x}{\partial t} \right)^2 + \left(\frac{\partial y}{\partial t} \right)^2 + \left(\frac{\partial z}{\partial t} \right)^2 \right]. \tag{7.25}$$

In terms of polar coordinates, for m_1 this becomes

$$T_1 = \frac{1}{2} m_1 r_1^2 \left[\left(\frac{\partial \theta}{\partial t} \right)^2 + \sin^2 \theta \left(\frac{\partial \phi}{\partial t} \right)^2 \right]. \tag{7.26}$$

When both atoms are included, the kinetic energy is given by

$$T = \frac{1}{2} \left(m_1 r_1^2 + m_2 r_2^2 \right) \left[\left(\frac{\partial \theta}{\partial t} \right)^2 + \sin^2 \theta \left(\frac{\partial \phi}{\partial t} \right)^2 \right]. \tag{7.27}$$

We can now write this equation using the moment of inertia, which gives

$$T = \frac{1}{2} I \left[\left(\frac{\partial \theta}{\partial t} \right)^2 + \sin^2 \theta \left(\frac{\partial \phi}{\partial t} \right)^2 \right]. \tag{7.28}$$

The derivatives are found from the conversions from Cartesian to polar coordinates. For example, the relationship for the x direction is

$$x = r \sin \theta \cos \phi, \tag{7.29}$$

so that

$$\frac{\partial x}{\partial t} = (r \cos \theta \cos \phi) \frac{\partial \theta}{\partial t} - (r \sin \theta \sin \phi) \frac{\partial \phi}{\partial t}. \tag{7.30}$$

The derivatives for the other variables are found similarly. Making use of the results, we can now write the operator for the kinetic energy as

$$\hat{T} = -\frac{\hbar^2}{2m} \left[\frac{1}{r^2} \frac{\partial}{\partial r} r^2 \frac{\partial}{\partial r} + \frac{1}{r^2 \sin \theta} \frac{\partial}{\partial \theta} \sin \theta \frac{\partial}{\partial \theta} + \frac{1}{r^2 \sin^2 \theta} \frac{\partial^2}{\partial \phi^2} \right]. \tag{7.31}$$

For a constant internuclear distance, the first term inside the brackets is 0. Furthermore, the kinetic energy must be described in terms of angular momenta in order to write the Hamiltonian operator. When this is done, we obtain

$$T = \frac{1}{2I}\left[p_\theta^2 + \frac{p_\phi^2}{\sin^2\theta}\right], \tag{7.32}$$

and the operators for the angular momenta are

$$\hat{p}_\theta = \frac{\hbar}{i}\frac{\partial}{\partial\theta} \quad \text{and} \quad \hat{p}_\phi = \frac{\hbar}{i}\frac{\partial}{\partial\phi}. \tag{7.33}$$

However, for rotation, where the moment of inertia is used, we assume no forces are acting on the rotor so that $V = 0$. Under these conditions, $H = T + V = T$ and the Hamiltonian operator is

$$\hat{H} = -\frac{\hbar^2}{2I}\left[\frac{1}{\sin\theta}\frac{\partial}{\partial\theta}\sin\theta\frac{\partial}{\partial\theta} + \frac{1}{\sin^2\theta}\frac{\partial^2}{\partial\phi^2}\right]. \tag{7.34}$$

The careful reader will observe that the form of the operator exactly replicates the angular portion of the Hamiltonian shown for the hydrogen atom in Eq. (4.12). Using this operator, the Schrödinger equation, $\hat{H}\psi = E\psi$, becomes

$$-\frac{\hbar^2}{2I}\left[\frac{1}{\sin\theta}\frac{\partial}{\partial\theta}\sin\theta\frac{\partial}{\partial\theta} + \frac{1}{\sin^2\theta}\frac{\partial^2}{\partial\phi^2}\right]\psi = E\psi. \tag{7.35}$$

It should also come as no surprise that the technique used in solving the equation is the separation of variables. Therefore, we assume a solution that can be written as

$$\psi(\theta,\phi) = Y(\theta,\phi) = \Theta(\theta)\,\Phi(\phi). \tag{7.36}$$

This product of two functions can be substituted into Eq. (7.35), and by dividing by the product, separating terms, and rearranging, we find that both parts of the equation are equal to some constant, which we choose to be $-m^2$. The two equations that are obtained by this separation are

$$\frac{d^2\Phi}{d\phi^2} = -m^2\Phi \tag{7.37}$$

$$\frac{1}{\sin\theta}\frac{d}{d\theta}\left(\sin\theta\frac{d\Theta}{d\theta}\right) - \frac{m^2}{\sin^2\theta}\Theta + \frac{2IE}{\hbar^2}\Theta = 0. \tag{7.38}$$

The equation in ϕ is of a form that has already been solved several times in earlier chapters. We can write the equation as

$$\frac{d^2\Phi}{d\phi^2} + m^2\Phi = 0. \tag{7.39}$$

The auxiliary equation can be written as

$$x^2 + m^2 = 0 \tag{7.40}$$

so that $x^2 = (-m^2)^{1/2}$ and $x = \pm im$. Therefore,

$$\Phi = e^{im\phi}, \tag{7.41}$$

but after a complete rotation through 2π rad the molecule has the same orientation. Therefore,

$$e^{im\phi} = e^{im(\phi+2\pi)}. \tag{7.42}$$

This is equivalent to saying that

$$e^{im2\pi} = 1. \tag{7.43}$$

Using Euler's formula,

$$e^{ix} = \cos x + i \sin x, \tag{7.44}$$

we find that

$$e^{im2\pi} = \cos 2\pi m + i \sin 2\pi m, \tag{7.45}$$

and this can be equal to 1 only when m is an integer, which means that $m = 0, \pm 1, \pm 2, \ldots$.

The second equation is simplified by letting $2IE/\hbar^2$ be equal to $l(l+1)$. Thus, we obtain

$$\frac{1}{\sin\theta}\frac{d}{d\theta}\sin\theta\frac{d}{d\theta}\Theta - \frac{m^2}{\sin^2\theta}\Theta + l(l+1)\Theta = 0. \tag{7.46}$$

This equation can be altered by taking the derivatives and collecting terms to give

$$\frac{d^2\Theta}{d\theta^2} + \frac{\cos\theta}{\sin\theta}\frac{d\Theta}{d\theta} + \left[l(l+1) - \frac{m^2}{\sin^2\theta}\right]\Theta = 0. \tag{7.47}$$

A transformation of variable from θ to x is accomplished by the following changes:

$$x = \cos\theta \tag{7.48}$$

$$\sin^2\theta = 1 - x^2 \tag{7.49}$$

$$\frac{dx}{d\theta} = -\sin\theta \tag{7.50}$$

$$\frac{d}{d\theta} = \frac{d\theta}{dx}\frac{d}{d\theta} = -\sin\theta\frac{d}{dx} \tag{7.51}$$

$$\frac{d^2}{d\theta^2} = \frac{d}{d\theta}\left[-\sin\theta\frac{d}{dx}\right] = -\cos\theta\frac{d}{dx} - \sin\theta\frac{d}{d\theta}\frac{d}{dx}. \tag{7.52}$$

Substituting for $d/d\theta$, the last relationship gives

$$\frac{d^2}{d\theta^2} = \sin^2\theta \frac{d^2}{dx^2} - \cos\theta \frac{d}{dx}. \tag{7.53}$$

Substituting these quantities into Eq. (7.47), we obtain

$$\frac{d^2\Theta(\theta)}{d\theta^2} + \frac{\cos\theta}{\sin\theta}\frac{d\Theta(\theta)}{d\theta} + \left[l(l+1) - \frac{m^2}{1-x^2}\right]\Theta(\theta) = 0, \tag{7.54}$$

which can be written as

$$\begin{aligned} \sin^2\theta \frac{d^2\Theta(x)}{dx^2} \quad - \quad & \cos x \frac{d\Theta(x)}{dx} + \frac{\cos\theta}{\sin\theta}\frac{d\Theta(\theta)}{d\theta} \\ + \quad & \left[l(l+1) - \frac{m^2}{1-x^2}\right]\Theta(x) = 0. \end{aligned} \tag{7.55}$$

Now, by replacing $d\Theta(\theta)/d\theta$ with $-\sin\theta d\Theta(\theta)/dx$ and $\sin^2\theta$ with $1-x^2$ we obtain

$$\left(1-x^2\right)\frac{d^2\Theta(x)}{dx^2} - 2x\frac{d\Theta(x)}{dx} + \left[l(l+1) - \frac{m^2}{1-x^2}\right]\Theta(x) = 0. \tag{7.56}$$

This equation has the form shown in Table 6.1 for Legendre's equation. Once again we have successfully reduced a problem in quantum mechanics to one of the famous differential equations shown in Table 6.1. The solution of the equation for the radial portion of the hydrogen atom problem was equivalent to solving Laguerre's equation. The angular portion of the hydrogen atom problem involved Legendre's equation, and the solution of the harmonic oscillator problem required the solution of Hermite's equation.

A series solution for Legendre's equation is well known and requires the series of polynomials known as the Legendre polynomials, which can be written as

$$P_l^{|m|}(\cos\theta). \tag{7.57}$$

Therefore, the wave functions for the rigid rotor are written as

$$\psi_{l,m}(\theta,\phi) = N P_l^{|m|}(\cos\theta)\, e^{im\phi}, \tag{7.58}$$

where N is a normalization constant. The solutions $\psi_{l,m}(\theta, \phi)$ are known as the spherical harmonics that were first encountered in this book in the solution of the wave equation for the hydrogen atom. It can be shown that the normalization constant can be written in terms of l and m as

$$N = \left[\frac{(2l+1)(l-|m|)!}{4\pi(l+|m|)!} \right]^{1/2} \tag{7.59}$$

so that the complete solutions can be written as

$$\psi_{l,m} = \left[\frac{(2l+1)(l-|m|)!}{4\pi(l+|m|)!} \right]^{1/2} P_l^{|m|}(\cos\theta)\, e^{im\phi}. \tag{7.60}$$

As we have seen, the solution to several quantum mechanical problems involves a rather heavy investment in mathematics, especially the solution of several famous differential equations. Although the complete details have not been presented in this book, the procedures have been outlined in sufficient detail so that the reader has an appreciation of the methods adequate for quantum mechanics at this level. The references at the end of this chapter should be consulted for more detailed treatment of the problems.

By comparison of Eqs. (7.38) and (7.46) we see that

$$\frac{2IE}{\hbar^2} = l(l+1) \tag{7.61}$$

or

$$E = \frac{\hbar^2}{2I} l(l+1). \tag{7.62}$$

Although it's not proven here, the restrictions on the values of m and l by the Legendre polynomials require that l be a nonnegative integer, and in this case, $l = 0, 1, 2, \ldots$. For a rotating diatomic molecule, the rotational quantum number is usually expressed as J, so the energy levels are given as

$$E = \frac{\hbar^2}{2I} J(J+1). \tag{7.63}$$

Therefore, the allowed rotational energies are

$$E_0 = 0; \quad E_1 = 2\frac{\hbar^2}{2I}; \quad E_2 = 6\frac{\hbar^2}{2I}; \quad E_3 = 12\frac{\hbar^2}{2I}; \text{ etc.}$$

Figure 7.3 shows the first few rotational energy levels.

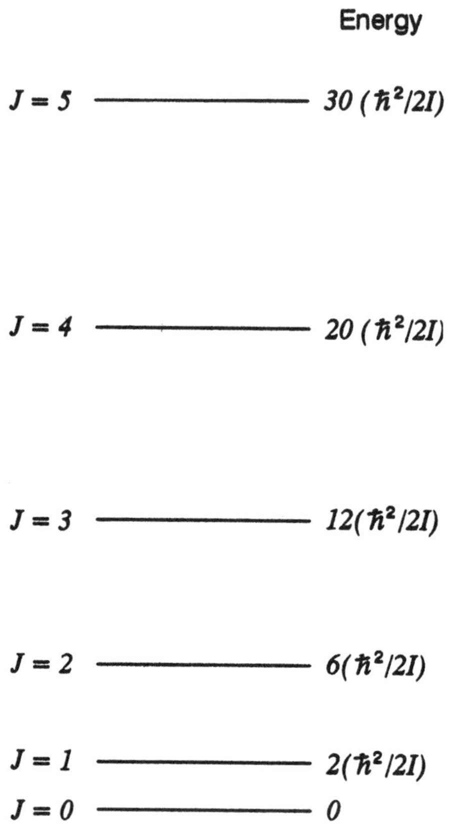

	Energy
$J = 5$	$30\,(\hbar^2/2I)$
$J = 4$	$20\,(\hbar^2/2I)$
$J = 3$	$12(\hbar^2/2I)$
$J = 2$	$6(\hbar^2/2I)$
$J = 1$	$2(\hbar^2/2I)$
$J = 0$	0

Figure 7.3 ▶ Rotational energies for a diatomic molecule (drawn to scale).

7.3 Heat Capacities of Gases

When gaseous molecules absorb heat, they undergo changes in rotational energies. Therefore, studying the thermal behavior of gases provides information on rotational states of molecules. It can be shown that for an ideal gas,

$$PV = nRT = \left(\frac{2}{3}\right) E, \tag{7.64}$$

where E is the total kinetic energy. Therefore, we can write

$$\frac{E}{T} = \frac{3}{2} nR. \tag{7.65}$$

TABLE 7.1 ▶ Heat Capacities of Gases at 25°C

Gas	C_v (cal/mol deg)	C_v (J/mol deg)
Helium	2.98	12.5
Argon	2.98	12.5
Hydrogen	4.91	20.5
Oxygen	5.05	21.1
Nitrogen	4.95	20.7
Chlorine	6.14	25.7
Ethane	10.65	44.6

If n is 1 mol and we change the temperature of the gas by 1 K there will be a corresponding change in E, which we will write as ΔE. Therefore,

$$\frac{\Delta E}{\Delta T} = \frac{3}{2}R. \tag{7.66}$$

The amount of heat needed to raise the temperature of 1 g of some material by 1 deg is the *specific heat* of the material. The *molar* quantity is called the *heat capacity* and is measured in J/mol K or cal/mol K. Equation (7.66) shows that the heat capacity of an ideal gas should be $(\frac{3}{2})R$, which is 12.47 J/mol K or 2.98 cal/mol K. Table 7.1 shows the heat capacities of several gases at constant volume. There are two different heat capacities in use, C_v the heat capacity at constant volume, and C_p, the heat capacity at constant pressure. It can be shown that $C_p = C_v + R$. If the gas is at constant pressure, heating the gas by 1 K causes an expansion of the gas, which requires work to push back the surroundings containing the gas. Therefore, the heat capacity at constant pressure is greater than the heat capacity at constant volume where the absorbed heat changes only the kinetic energy of the gas.

The experimental heat capacities for helium and argon are identical to those predicted by the ideal gas equation (12.5 J/mol K). However, for all of the other gases listed in the table, the values do not agree with the ideal gas heat capacity. At first glance, it appears that the monatomic gases have heat capacities that agree with the simple model based on increasing the kinetic energy of the molecules while gases consisting of diatomic and polyatomic molecules do not.

The motion of gaseous molecules through space is described in terms of motion in three directions. Since absorbed heat increases the kinetic energy in three directions, on average, the same amount of heat goes into increasing the energy in each direction. That is, $(\frac{3}{2})R = 3(\frac{1}{2})R$, with $(\frac{1}{2})R$ going to increase kinetic energy in each direction. Each direction is called a *degree of freedom* and the overall kinetic energy is the sum of the energy in each direction. However, this is not the only way in which the heat is absorbed by molecules if they consist of more than one atom. There are other degrees of freedom in addition to linear motion through space (*translation*). The other ways in which molecules absorb heat are by changing *rotational* and *vibrational* energies.

A principle known as the Law of Equipartition of Energy can be stated quite simply as follows: *If a molecule can absorb energy in more than one way, it can absorb equal amounts in each way*. This is also true for translation where $(\frac{1}{2})R$ absorbed goes toward increasing the kinetic energy in each of the three directions. In order to use this principle, we must know the number of "ways" (degrees of freedom) a molecule can rotate and vibrate.

As will be shown later, for a diatomic molecule (which is linear) there are only two degrees of rotational freedom, and we can infer that this would also be true for other *linear* molecules. In the case of linear molecules for which absorbed heat can change their rotational energy, there will be $2(\frac{1}{2})R$ absorbed. The total heat absorbed for a mole of a gas composed of linear molecules will be

$$3\left(\frac{1}{2}\right)R = \left(\frac{3}{2}\right)R = 12.5 \text{ J/molK (translation)}$$

$$2\left(\frac{1}{2}\right)R = R = 8.3 \text{ J/molK (rotation)}$$

for a total heat capacity of 20.8 J/mol K, which is equal to $(\frac{5}{2})R$. Note that this value is very close to the actual heat capacities shown in Table 7.1 for H_2, O_2, and N_2. It is reasonable to conclude that for these diatomic molecules, absorbed heat is changing only their translational and rotational energies. We should note that for nonlinear molecules, there are three degrees of rotational freedom, each of which can absorb $(\frac{1}{2})R$.

Although we have explained the value of $(\frac{5}{2})R$ for the heat capacity of diatomic molecules at 25°C, we should note that at very high temperatures (1500 K), the heat capacity of hydrogen is about 29.2 J/mol K. This shows that at high temperature the H_2 molecule can absorb energy in some way other than changing its translational and rotational energies. The additional means by which H_2 molecules can absorb energy is by changing vibrational energy. A chemical bond is not totally rigid and in a diatomic molecule the bond can be represented as a spring (see Figure 7.4). However, the vibrational energy (as well as the rotational energy) is quantized (see Sections 6.6 and 7.2). Since the observed heat capacity of H_2 at 25°C can be accounted for in terms of changes in only translational and

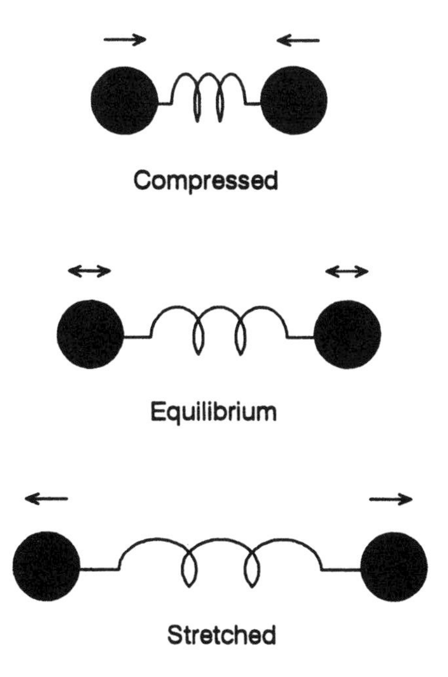

Figure 7.4 ▶ Motion of a diatomic molecule during vibration.

rotational energies, we conclude that the molecules cannot change vibrational energy at this low temperature. Therefore, it should be apparent that the rotational energy states must be separated by a energy smaller than that which separates vibrational energy levels. We will now explore the nature of rotational and vibrational energy states for gaseous diatomic molecules in greater detail.

7.4 Energy Levels in Gaseous Atoms and Molecules

Emission spectra for atoms appear as a series of lines since electrons fall from higher energy states to lower ones and emit energy as electromagnetic radiation. You should recall the line spectrum of hydrogen (see Chapter 1) and the fact that the Lyman series is in the ultraviolet (UV) region and the Balmer series is in the visible region. Consequently, spectroscopy carried out to observe the transitions between the electronic energy levels often involves radiation in the visible and ultraviolet regions of the electromagnetic spectrum.

For a spectral line of 6000 Å (600 nm), which is in the visible light region, the corresponding energy is

$$E = h\nu = \frac{hc}{\lambda} = \frac{(6.63 \times 10^{-27} \text{ erg s} \times (3.00 \times 10^{10} \text{ cm/s})}{6.00 \times 10^{-5} \text{ cm}}$$
$$= 3.3 \times 10^{-12} \text{ erg.}$$

Now, converting this to a molar quantity by multiplying by Avogadro's number gives 2.0×10^{12} erg/mol, and converting to kilojoules gives an energy of about 200 kJ/mol:

$$E = \frac{(3.3 \times 10^{-12} \text{ erg/molecule}) \times (6.02 \times 10^{23} \text{ molecules/mol})}{10^{10} \text{ erg/kJ}}$$
$$= 200 \text{ kJ/mol.}$$

This is within the typical range of energies separating electronic states, which is about 200–400 kJ/mol.

Although we have been discussing electronic energy levels in atoms, the electronic states in molecules are separated by similar energies. In general, the electromagnetic radiation *emitted* from atoms is usually studied, but it is the radiation *absorbed* by molecules that is usually examined by UV/visible spectroscopy. In addition to electronic energy levels, molecules also have vibrational and rotational energy states. As shown in Figure 7.4, the atoms in a diatomic molecule can be viewed as if they are held together by bonds that have some stretching and bending (vibrational) capability, and the whole molecule can rotate as a unit. Figure 7.5 shows the relationship between the bond length and the potential energy for a vibrating molecule. The bottom of the potential well is rather closely approximated by a parabolic potential (see Chapter 6).

The difference in energy between adjacent vibrational levels ranges from about 10 to 40 kJ/mol. Consequently, the differences in energy between two vibrational levels correspond to radiation in the infrared region of the spectrum. Rotational energies of molecules are also quantized, but the difference between adjacent levels is only about 10–40 J/mol. These small energy differences correspond to electromagnetic radiation in the far-infrared (or in some cases the microwave) region of the spectrum. Therefore, an infrared spectrometer is needed to study changes in vibrational or rotational states in molecules. The experimental technique known as infrared spectroscopy is concerned with changes in vibrational and rotational energy levels in molecules. Figure 7.6 shows the relationship between the electronic, vibrational, and rotational energy levels for molecules and the approximate range of energy for each type of level.

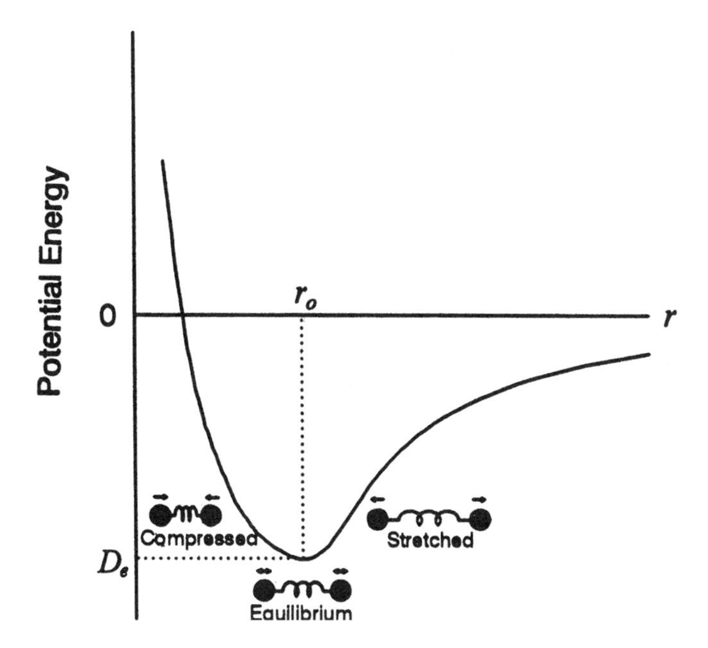

Figure 7.5 ▶ Potential energy versus bond length for a diatomic molecule.

We saw clearly in the study of heat capacities of gases how the existence of rotational states affects the heat capacity. For He the heat capacity is $(\frac{3}{2})R$; that for H_2 is $(\frac{5}{2})R$ at room temperature but approaches $(\frac{7}{2})R$ at high temperatures. The reason for this is that for H_2, the absorbed energy not only can change the kinetic energies (translation) of the molecules, but it also can change their rotational energies. At quite high temperatures, the vibrational energies of the molecules can also change. The Law of Equipartition of Energy states that if a molecule can absorb energy in more than one way, it can absorb equal amounts in each way. Since there are three degrees of translational freedom [three components (x, y, and z) to velocity or kinetic energy], each degree of freedom can absorb $(\frac{1}{2})R$ as the molecules change their kinetic energies. Because H_2 (and all other diatomic molecules) are linear, there are only two degrees of rotational freedom, as shown in Figure 7.7. If we examine the possibility of rotation around the z axis (the internuclear axis), we find that the moment of inertia is extremely small. Using a nuclear radius of 10^{-13} cm, the value for I is several orders of magnitude larger for rotation around the x and y axes since the bond length is on the order of 10^{-8} cm. The energy of rotation is given by

$$E = \frac{\hbar^2}{2I}, \tag{7.67}$$

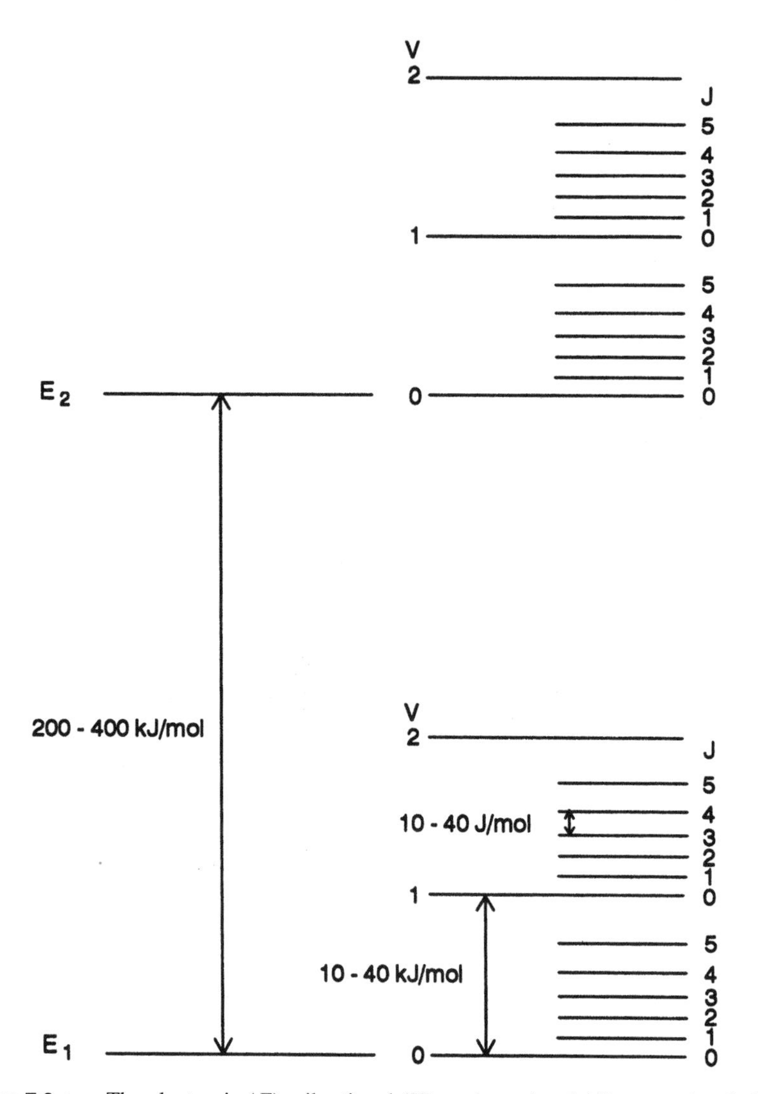

Figure 7.6 ▶ The electronic (E), vibrational (V), and rotational (J) energy levels for a diatomic molecule with typical ranges of energies.

and we see that the difference between two adjacent rotational levels ($J = 0$ to $J = 1$),

$$E = \frac{h^2}{8\pi^2 I}, \tag{7.68}$$

is very large for rotation around the z axis. Therefore, an increase in rotational energy around the internuclear axis for linear molecules (the z axis

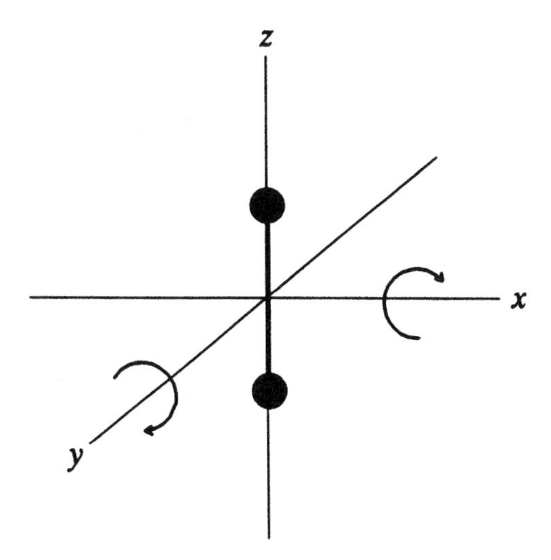

Figure 7.7 ▶ The rotational degrees of freedom for a diatomic molecule.

because that is the axis of highest symmetry) is not observed and linear molecules have only two degrees of rotational freedom.

There is only one degree of vibrational freedom, but it counts double $[2(\frac{1}{2})R]$ since it involves an increase in both kinetic and potential energy. Therefore, when H_2 is absorbing heat in *all* possible ways the heat capacity is given as the sum $(\frac{3}{2})R + (\frac{2}{2})R + (\frac{2}{2})R = (\frac{7}{2})R$, the value at high temperatures.

At room temperature (about 300 K), RT is the thermal energy available and it is calculated from

$$(8.3144 \text{ J/mol K}) \times (300 \text{ K}) = 2500 \text{ J/mol}.$$

Therefore, the very large separation between electronic states (perhaps 100 kJ/mol) means that only the lowest electronic state will be populated. Likewise, for most molecules the difference of 10–40 kJ/mol between vibrational states means that only the lowest vibrational state will be populated at low temperatures. The relatively small differences between rotational states makes it possible for several rotational states to be populated (unequally) even at room temperature.

7.5 Molecular Spectra

We have described the energy levels of molecules, and the differences between these states were shown to correspond to different regions of the electromagnetic spectrum. Electronic energy levels are usually separated by sufficient energy to correspond to radiation in the visible and ultraviolet regions of the spectrum. Vibrations have energies of such magnitudes that the changes in energy levels are associated with the infrared region of the spectrum. Therefore, infrared spectroscopy deals essentially with the changes in vibrational energy levels in molecules. The energy level diagram shown in Figure 7.6 shows that the rotational levels are much more closely spaced than are the vibrational energy levels. Accordingly, it is easy to produce a change in the rotational state of a small molecule as the vibrational energy is being changed. In fact, for some molecules, there is a restriction (known as a *selection rule*) that permits the vibrational state of the molecule to change only if the rotational state is changed as well. Such a molecule is HCl. Therefore, if one studies the infrared spectrum of gaseous HCl, a series of peaks that corresponds to the absorption of the infrared radiation as the rotational energy level changes at the same time as the vibrational state changes is seen. There is *not* one single absorption peak due to the change in vibrational state, but a series of peaks as the vibrational state and rotational states change. Figure 7.8 shows how these transitions are related. As shown in Figure 7.8, the quantized vibrational states are characterized by the quantum number V, while the rotational states are identified by the quantum number J.

For a diatomic molecule, the rotational energies are determined by the masses of the atoms and the distance of their separation. Therefore, from the experimentally determined energies separating the rotational states it is possible to calculate the distance of separation of the atoms if their masses are known. However, if a molecule rotates with a higher rotational energy, the bond length will be slightly longer because of the centrifugal force caused by the rotation. As a result of this, the spacing between rotational energy states for $J = 1$ and $J = 2$ is slightly different than it is for $J = 4$ and $J = 5$. Figure 7.8 shows this effect graphically where the difference between adjacent rotational states increases slightly at higher J values.

As molecules change in vibrational energy states, the selection rule specifies that they must also change their rotational state. However, since several rotational states are populated, some of the molecules will increase in rotational energy and some will decrease in rotational energy as all of the molecules increase in vibrational energy. The rotational states are designated by a quantum number J so that the levels are $J_0, J_1, J_2, \ldots,$ and the vibrational energy states are denoted by the vibrational quantum numbers, V_0, V_1, V_2, etc. For the transition in which the vibrational energy is increasing from $V = 0$ to $V = 1$, $\Delta V = +1$, but ΔJ can be $+1$ or -1, depending on whether the molecules are increasing in rotational

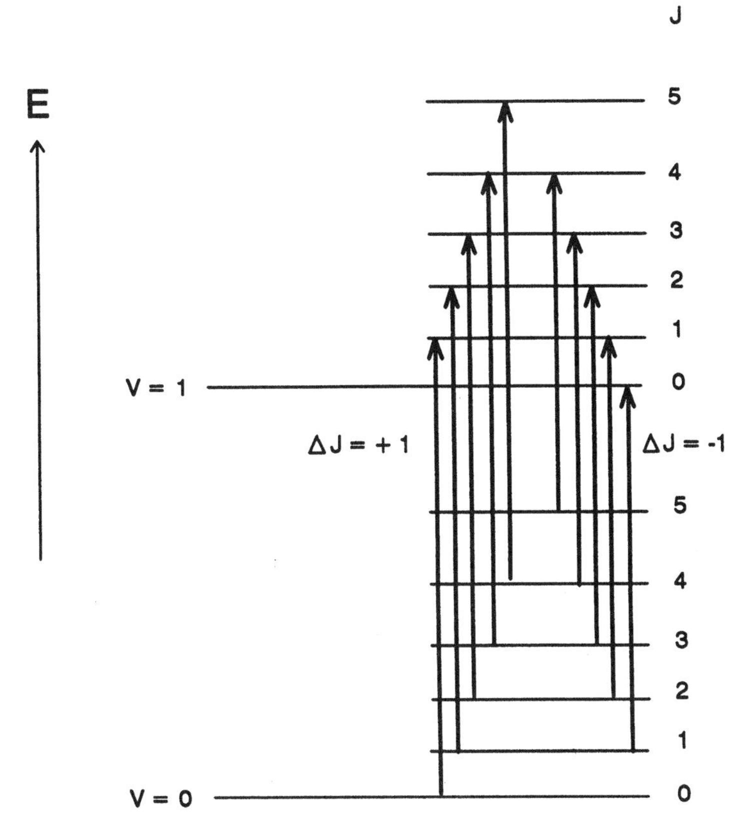

Figure 7.8 ▶ Changes in energy as molecules are excited from the lowest vibrational state to the next higher one. All of the molecules are increasing in vibrational energy, but where $\Delta J = +1$ the molecules are increasing in rotational energy and where $\Delta J = -1$ the molecules are decreasing in rotational energy.

energy ($\Delta J = +1$) or decreasing in rotational energy ($\Delta J = -1$). Figure 7.9 shows the vibration–rotation spectrum for gaseous HCl under moderately high resolution obtained by the author in 1969. The spectrum shows a series of sharp peaks appearing in two portions. All of the molecules are increasing in vibrational energy, but some are increasing in rotational energy and some are decreasing in rotational energy. Note that there appears to be a gap in the middle of the spectrum where a peak is missing. The missing peak represents the transition between the first two vibrational states with no change in rotational state. Since this type of transition is prohibited for the HCl molecule, that peak is missing.

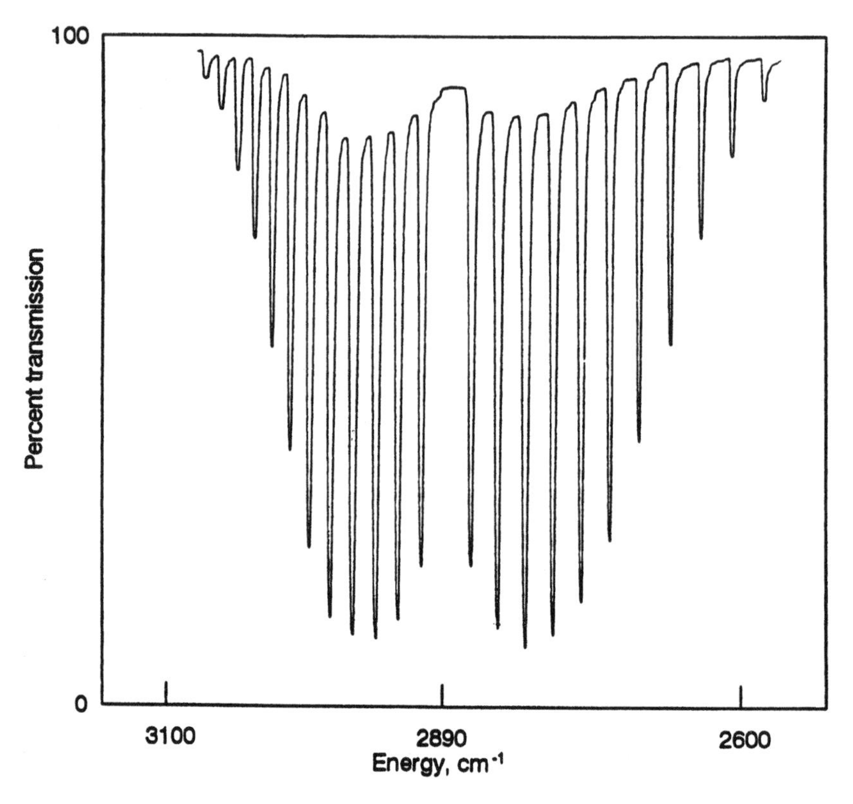

Figure 7.9 ▶ The vibration–rotation spectrum of gaseous HCI. The spacing between adjacent peaks is about 20.7 cm^{-1}.

If a photograph is taken using a camera with a poor lens, details of the subject are not visible. Closely spaced lines appear as a blur. The same subject photographed with a camera having a lens of high quality will show much better resolution so that small details are visible. A similar situation exists with spectra. If a spectrometer having poor resolution is used to record the spectrum of gaseous HCl, the individual sharp peaks are not resolved and only two large peaks are observed. With a better spectrometer, all of the peaks are resolved (as shown in Figure 7.9) and the two series of sharp peaks are observed. A spectrum such as that in Figure 7.9 showing the absorption of infrared radiation as the molecules change vibrational and rotational state is said to show *rotational fine structure*. If ultrahigh resolution is used, the individual sharp peaks are seen to split into two smaller peaks. The reason for this is that chlorine exists as a mixture of ^{35}Cl and ^{37}Cl and the rotational energy of HCl depends on the masses of

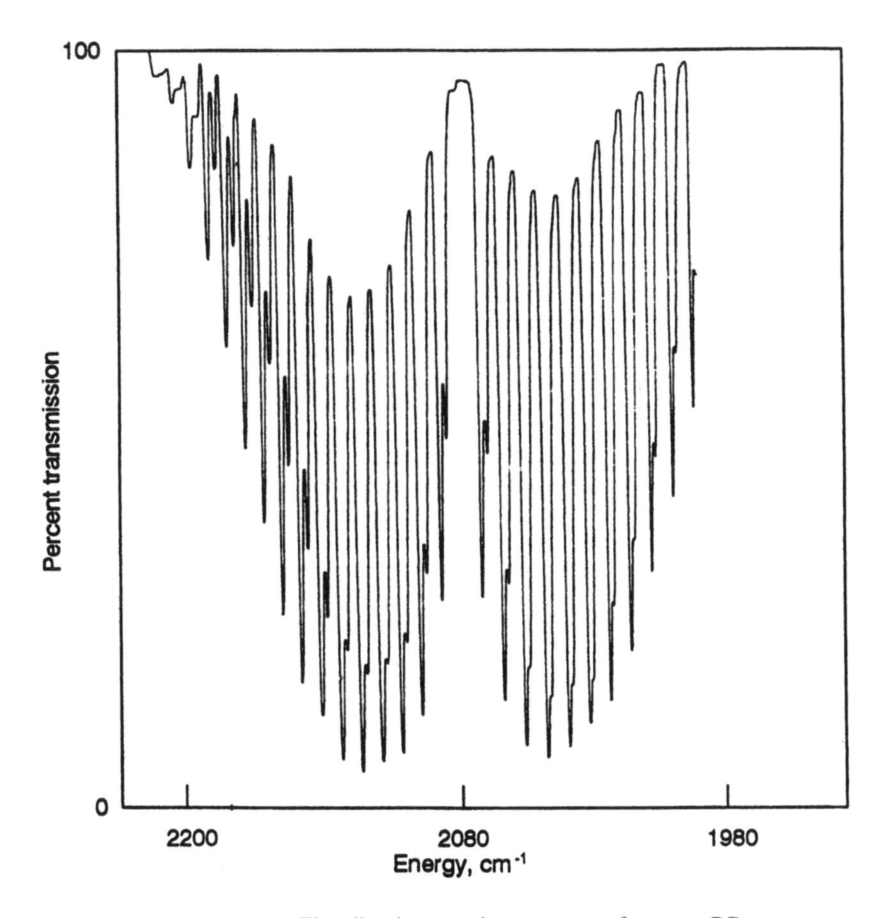

Figure 7.10 ▸ The vibration–rotation spectrum of gaseous DBr.

the atoms. Therefore, $H^{35}Cl$ and $H^{37}Cl$ have slightly different rotational energies, which causes each peak to be split into two closely spaced peaks. Figure 7.10 shows how the peaks split for gaseous DBr. The reason for the peaks splitting in this case is that bromine occurs naturally as two isotopes, ^{79}Br (mass 78.92) and ^{81}Br (mass 80.92). The moments of inertia are slightly different for $D^{79}Br$ and $D^{81}Br$, resulting in rotational states that are slightly different.

We will now show how to obtain molecular parameters from the spectrum shown in Figure 7.9, which represents the transition of HCl from the lowest vibrational state to the first excited vibrational state with rotational fine structure. Although the transition corresponding to $V = 0$ to $V = 1$ with $\Delta J = 0$ is missing, it would occur at about 2885 cm^{-1}. We can use that information to determine the force constant for the H–Cl bond. The frequency is related to the force constant by

$$\nu = \frac{1}{2\pi}\sqrt{\frac{k}{\mu}}, \tag{7.69}$$

and the frequency is related to $\bar{\nu}$ by $\nu = c/\lambda = c\bar{\nu}$. The reduced mass, μ, is $m_{\mathrm{H}}m_{\mathrm{Cl}}/(m_{\mathrm{H}} + m_{\mathrm{Cl}}) = 1.63 \times 10^{-24}$ g. Therefore,

$$k = \mu(2\pi c\bar{\nu})^2 = 4.82 \times 10^5 \text{ dyn/cm}. \tag{7.70}$$

A more commonly used unit for the force constant is mdyn/Å. Because 1 dyn $= 10^3$ mdyn and 1 cm $= 10^8$ Å, the force constant in these units is 4.82 mdyn/Å.

The spacing between rotational bands is 20.7 cm^{-1}, and from that we can determine the internuclear distance for HCl. The rotational energy can be represented as

$$E = \frac{\hbar^2}{2I}J(J+1) = \frac{h^2}{8\pi^2 I}J(J+1), \tag{7.71}$$

and for the transition $J = 0$ to $J = 1$,

$$\Delta E = E_1 - E_0 = \frac{h^2}{4\pi^2 I} = 20.7 \text{ cm}^{-1}. \tag{7.72}$$

This energy can be converted into ergs,

$$\Delta E = h\nu = hc/\lambda = hc\bar{\nu} \tag{7.73}$$

$$\Delta E = \left(6.63 \times 10^{-27} \text{ erg s}\right) \times \left(3.00 \times 10^{10} \text{ cm/s}\right) \times 20.7 \text{ cm}^{-1}$$
$$= 4.12 \times 10^{-15} \text{ erg}.$$

Therefore, the moment of inertia is

$$I = \frac{h^2}{4\pi^2(\Delta E)} = 2.70 \times 10^{-40} \text{ g cm}^2. \tag{7.74}$$

Since $I = \mu R^2$, we can solve for the internuclear distance, R, and find that it is 1.29×10^{-8} cm $= 1.29$ Å $= 129$ pm. These simple applications show the utility of infrared spectroscopy in determining molecular parameters.

7.6 Structure Determination

For the simplest type of molecule, a diatomic molecule, there is only one vibration possible. It is the stretching of the chemical bond. That bond has an energy related to the distance between the atoms. Figure 7.5 shows that type of relationship in which there is a lowest energy (D_e), which occurs at the normal bond distance, r_0. When the bond is either longer or shorter than this distance, the energy is higher and the bond is less stable.

In the case of more complicated polyatomic molecules, there is a potential energy curve for each type of bond between atoms in the molecule. Therefore, changes in the vibrational levels of these bonds result in absorptions in the spectrum characteristic of the types of bonds present. It is thus possible in many cases to attribute absorption bands in the infrared spectrum to the types of bonds present in the molecule and to determine a great deal about how the atoms are arranged. This application of infrared spectroscopy is of tremendous importance to the practicing chemist. With an elemental analysis to determine the ratio of atoms present (empirical formula), a molecular weight to determine the actual numbers of atoms present, and an infrared spectrum to identify the kinds of bonds present, a chemist is well along toward identifying a compound.

For relatively simple molecules, it is possible to determine the structure by infrared spectroscopy. The total number of fundamental vibrations for a molecule having N atoms is $3N - 5$ if the molecule is linear and $3N - 6$ if the molecule is nonlinear. Thus, for a diatomic molecule, the total number of vibrations is $3N - 5 = 3 \times 2 - 5 = 1$. For a triatomic molecule, there will be $3N - 6 = 3 \times 3 - 6 = 3$ vibrations for an angular or bent structure and $3N - 5 = 3 \times 3 - 5 = 4$ vibrations if the structure is linear. For molecules consisting of three atoms, it is possible to determine the molecular structure on the basis of the number of vibrations that lead to the absorption of energy.

For the molecule SF_6, $N = 7$ so there will be $3 \times 7 - 6 = 15$ fundamental vibrations. Each vibration has a set of vibrational energy levels similar to those shown in Figure 7.6. Consequently, the changes in vibrational energy for each type of vibration will take place with different energies, and the bands can sometimes be resolved and assigned. However, not all of these vibrations lead to absorption of infrared (IR) radiation. There are also other bands called overtones and combination bands. Therefore, the total number of vibrations is large, and it is not likely that the molecular structure could be determined solely on the basis of the number of absorption bands seen in the IR spectrum.

For a change in vibrational energy to be observed as an absorption of electromagnetic radiation (called an *IR active* change), the change in vibrational energy state must result in a change in dipole moment. This is because electromagnetic radiation consists of an oscillating electric and magnetic field (see Chapter 1). Therefore, an electric dipole can interact with the radiation and absorb energy to produce changes in the molecule. If we consider HCl as an example, there is a single vibration, the stretching of the H–Cl bond. We know that the dipole moment (μ) is the product of the amount of charge separated (q) and the distance of separation (r):

$$\mu = q \cdot r. \tag{7.75}$$

When HCl is excited from the lowest vibrational energy to the next higher one, its average bond length, r, increases slightly. Since the dipole moment depends on r, the dipole moment of HCl is slightly different in the two vibrational states. Therefore, it is possible to observe the change in vibrational states for HCl as an absorption of infrared radiation. For the hydrogen molecule, H–H, there is no charge separation so $\mu = 0$. Increasing the bond length does not change the dipole moment, so a change in vibrational energy of H_2 cannot be seen as an absorption of infrared radiation. This called an *IR inactive* vibrational change.

Let us consider a molecule like CO_2. It is reasonable to assume that the atoms are arranged in either a linear or a bent (angular) structure. Table 7.2 shows the types of vibrations that would be possible for these two structures. Since the two bending modes of the linear structure are identical

TABLE 7.2 ▶ Vibrations Possible for Assumed Linear and Bent Structures for the Carbon Dioxide Molecule

Assumed linear structure		Assumed bent structure	
← → O–C–O	Symmetric stretch, no change in dipole moment, IR inactive	Symmetric stretch, change in dipole moment, IR active	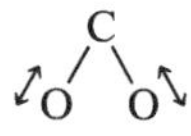 C ↙O O↘ (with C–O bonds)
→ → O–C–O	Asymmetric stretch, change in dipole moment, IR active	Asymmetric stretch, change in dipole moment, IR active	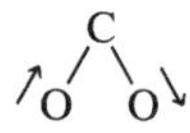 C ↗O O↘ (with C–O bonds)
↻ O–C–O ↺ O–C–O + − +	Bending, two modes, change in dipole moment, IR active	Bending, change in dipole moment, IR active	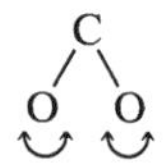 C O O ↻ ↻

Note. The + and − signs are used to denote motion perpendicular to the plane of the page.

except for being perpendicular to each other, they involve the same energy. Accordingly, only one vibrational absorption band should be seen in the IR spectrum for bending in either direction. If CO_2 were linear, a change in the symmetric stretching vibration would not cause a change in dipole moment since the effects on each C–O bond would exactly cancel. Changing the energy level for the asymmetrical stretching vibration does cause a change in dipole moment (the effects in opposite directions do not cancel) and one vibrational band is seen corresponding to asymmetric stretching. Therefore, if CO_2 were linear, there would be two vibrational bands in the infrared spectrum.

If CO_2 has an angular structure, both of the stretching vibrations would cause a change in dipole moment because the effects on the C–O bonds do not cancel. Thus, the changes in both the symmetric and the asymmetric stretching vibrations would be IR active. The bending vibration for the angular structure would also cause a change in dipole moment because the C–O bonds are not directly opposing each other. Consequently, if the structure were angular, there would be three bands in the vibrational region of the infrared spectrum. The actual spectrum of CO_2 shows two bands for vibrational changes at 2350 and 667 cm^{-1}. Therefore, the structure of CO_2 must be linear. On the other hand, NO_2 shows vibrational bands at 1616, 1323, and 750 cm^{-1}, indicating that it has a bent structure.

For molecules that contain a large number of atoms, it may become difficult if not impossible to sort out the bands to establish structure on this basis alone. The techniques that we have discussed are one basis for *how* we know that some of the molecular structures described in chemistry textbooks are correct.

7.7 Types of Bonds Present

One of the convenient aspects of vibrational changes in chemical bonds is that for a particular type of bond the remainder of the molecule has a relatively small effect on the vibrational energy levels. For example, the change in stretching vibrational energy of the –O–H bond requires about the same energy regardless of what is bonded on the other side of the oxygen atom. Accordingly, H_3C–O–H (usually written as CH_3OH) and C_2H_5–O–H give absorptions of energy in the same region of the electromagnetic spectrum, at about 3600 cm^{-1} or at a wavelength of 2780 nm (2.78×10^{-4} cm). Therefore, if the infrared spectrum of a compound exhibits an absorption band at this position, we can be reasonably sure that the molecules contain O–H

TABLE 7.3 ▶ Types of Bonds and Typical Energies They Absorb

Bond	Approximate absorption region (cm^{-1})
O–H	3600
N–H	3300
C–H	3000
C=O	1700
C–Cl	700
C≡N	2100
C–C	1000
C=C	1650
C≡C	2050

bonds. A few common types of bonds and the typical regions where they absorb energy as they change stretching vibrations are given in Table 7.3.

We can see how the units cm^{-1} come into use in the following way: Remember that $E = h\nu$ and $\lambda\nu = c$. Therefore,

$$E = h\nu = hc/\lambda. \tag{7.76}$$

If we represent $1/\lambda$ as $\bar{\nu}$, then

$$E = hc\bar{\nu}. \tag{7.77}$$

The energy units for a single molecule work out as:

$$(\text{erg s}) \times (\text{cm/s}) \times (1/\text{cm}) = \text{erg}.$$

We can then convert erg/molecule to kJ/mol or kcal/mol.

Using a more extensive table of stretching frequencies, it is frequently possible to match the observed peaks in an infrared spectrum to the known values for the various types of bonds and thereby determine the types of bonds present in the compound. This information, along with percent composition and molecular weight, is frequently sufficient to identify the compound. Thus, infrared spectroscopy can be used in certain circumstances to determine bond lengths, molecular structure, and the types of bonds present in molecules. These applications make it one of the most useful tools for the study of materials by chemists, although many other experimental techniques [X-ray diffraction, nuclear magnetic resonance (NMR), electron spin resonance (ESR), etc.] are required for the complete study of matter. It is interesting to note, however, that the existence of rotational and vibrational levels in molecules is indicated by the study of a topic as basic as the heat capacities of gases!

References for Further Reading

- Drago, R. S. (1992). *Physical Methods for Chemists*, 2nd ed., Chap. 6. Saunders College Publishing, Philadelphia. This book is a monumental description of most of the important physical methods used in studying molecular structure and molecular interactions. It is arguably the most influential book of its type.
- Harris, D. C., and Bertolucci, M. D. (1989). *Symmetry and Spectroscopy*, Chap. 3. Dover, New York. Detailed treatment of all aspects of vibrational spectroscopy.
- Laidlaw, W. G. (1970). *Introduction to Quantum Concepts in Spectroscopy*. McGraw–Hill, New York. A good treatment of the quantum mechanical models that are useful for interpreting spectroscopy.
- Sonnessa, A. J. (1966). *Introduction to Molecular Spectroscopy*. Reinhold, New York. This book is probably hard to find now, but it is one of the best introductory books available on the subject of molecular spectroscopy.
- Wheatley, P. J. (1959). *The Determination of Molecular Structure*. Oxford Univ. Press, London. A classic introduction to experimental determination of molecular structure. Excellent discussion of spectroscopic methods.

Problems

1. The force constant for the C–H radical is 4.09×10^5 dyn/cm. What would be the wave number for the fundamental stretching vibration?

2. For HI, the bond length is about 1.60 Å or 160 pm. What would be the spacing between consecutive rotational bands in the IR spectrum of HI?

3. For CO, the change in rotational state from $J = 0$ to $J = 1$ gives rise to an absorption band at 0.261 cm, and that for $J = 1$ to $J = 2$ is associated with an absorption band at 0.522 cm. Use this information to determine the bond length of the CO molecule.

4. The vibrational–rotational spectrum shown in Figure 7.9 was obtained using an infrared spectrometer of limited resolution. Therefore, the bands do not show the separation that actually exists due to $H^{35}Cl$ and $H^{37}Cl$. What degree of resolution would the spectrometer need to have in order to show that separation? Assume that the bond lengths of the molecules are the same, 127.5 pm.

5. Figure 7.10 shows the vibration–rotation spectrum of DBr with the splitting caused by ^{79}Br and ^{81}Br. What would the splitting of the peaks caused by the difference in isotopic masses in $D^{35}C1$ and $D^{37}Cl$ be when measured in cm^{-1}?

6. The spacing between vibrational levels for HCl is about 2890 cm^{-1}, where the missing peak would occur in the spectrum shown in Figure 7.9. Calculate the force constant for the H–Cl bond in

  (a) dyn/cm,

  (b) mdyn/Å, and

  (c) N/m.

7. The spacing between rotational levels for HCl is about 20.7 cm^{-1}. Use this value to calculate the bond length in HCl.

8. For a rigid rotor, the rotational energy can be written as $L^2/2I$, where I is the moment of inertia and L is the angular momentum whose operator is

$$\hat{L} = \frac{\hbar}{i}\frac{\partial}{\partial \phi}.$$

Obtain the operator for rotational energy. Write the Schrödinger equation, and from the form of the equation, tell what functions could give acceptable solutions. Use the functions to evaluate the energy levels for rotation.

9. For CO, the bond length is 113 pm for both $^{12}CO^{16}O$ and $^{14}C^{16}O$.

  (a) Determine the moments of intertia for the two molecules.

  (b) Determine the difference in the energy between the $J = 1$ and $J = 2$ for $^{12}C^{16}O$ and $^{14}C^{16}O$.

10. Calculate the energy of the first three rotational states $H^{35}Cl$ and $H^{37}Cl$. The bond length is 129 pm.

▶ Chapter 8

Barrier Penetration

8.1 The Phenomenon of Barrier Penetration

One of the most interesting differences between classical and quantum descriptions of behavior concerns the phenomenon of barrier penetration or tunneling. Consider a particle approaching a barrier of height U_0 as shown in Figure 8.1. Suppose the particle has an energy E and the height of the barrier is such that $U_0 > E$. Classically, the particle cannot penetrate the barrier and does not have enough energy to get over the barrier. Therefore, if the particle has an energy less than U_0, it will be reflected by the barrier. If the particle has an energy greater than the height of the barrier, it can simply pass over the barrier. Because the particle with $E < U_0$ cannot get over the barrier and cannot penetrate the barrier, the regions inside the barrier and to the right of it are forbidden to the particle in the classical sense.

According to quantum mechanics, the particle (behaving as a wave) can not only penetrate the barrier, it can pass through it and appear on the other side! Since the moving particle has a wave character, it must have an amplitude and that amplitude is nonzero. The probability of finding the particle is proportional to the square of the amplitude function. Therefore, all regions are accessible to the particle even though the probability of it being in a given region may be very low. This is the result of the wave function for the moving particle not going to 0 at the boundary.

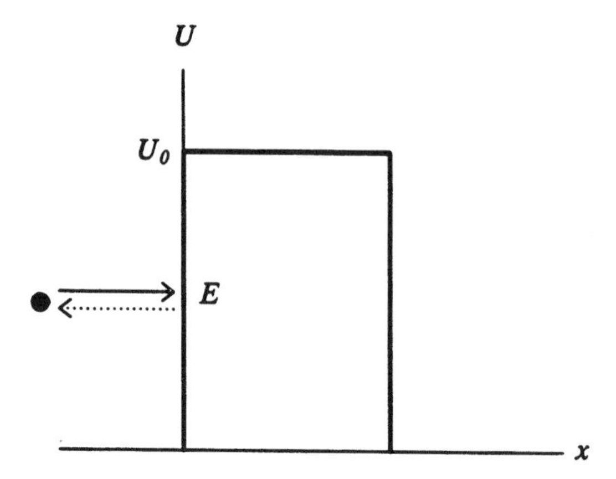

Figure 8.1 ► A particle with energy E colliding with an energy barrier.

8.2 The Wave Equations

In treating the barrier penetration problem by quantum mechanics, it must be recognized that the entire area surrounding the barrier is accessible to the particle. Since there are three regions (inside the barrier and to the left and right of it), there will be three wave equations to solve.

It is natural to try to determine the connection between the wave inside the barrier and on either side of it. The connection must be a smooth, continuous one given by the restrictions on the wave function and its first derivative (see Chapter 2). We can obtain the general form of the function by analogy to a simple example. Suppose light is shined into a solution that absorbs light in direct proportion to the intensity of the light (see Figure 8.2). Then, using k as the proportionality constant, we can write

$$-\frac{dI}{dx} = kI. \tag{8.1}$$

Rearrangement of Eq. (8.1) gives

$$-\frac{dI}{I} = k\,dx. \tag{8.2}$$

This equation can be integrated between limits of I_0 (the incident beam intensity) at distance 0 to some other intensity, I, after traveling a distance of x in the solution:

$$\ln \frac{I_0}{I} = kx. \tag{8.3}$$

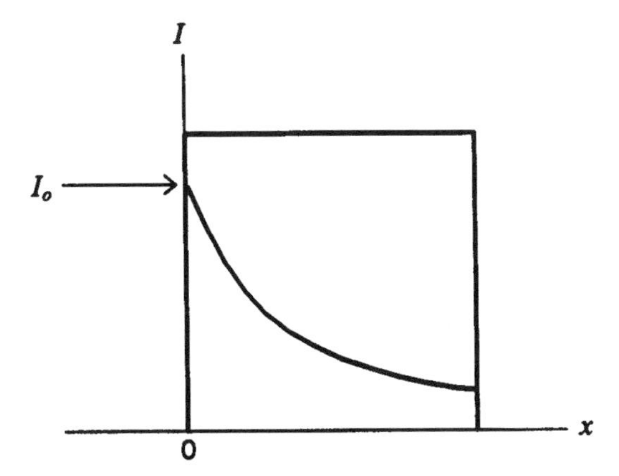

Figure 8.2 ▶ Variation of beam intensity with penetration depth.

By taking the antilogarithm of both sides of the equation, it can be written as

$$I = I_0 e^{-kx}. \tag{8.4}$$

From this equation, we see that the beam decreases in intensity in an exponential way with the distance it has penetrated into the solution. Therefore, we suspect that a smooth exponential decrease in the wave function should occur within the barrier, and that is, in fact, the case. Figure 8.3 shows the barrier and defines the parameters used in solving the problem of barrier penetration.

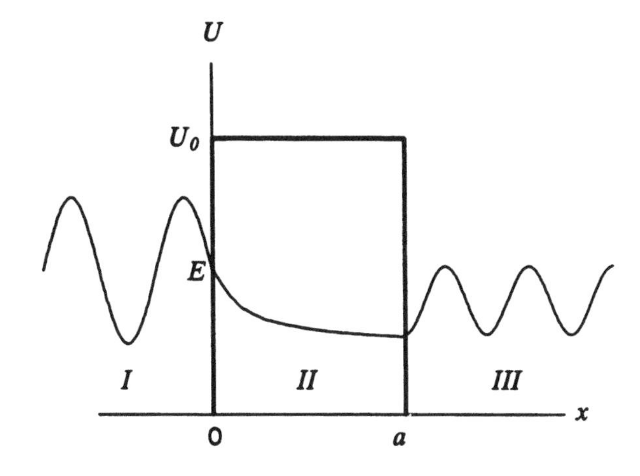

Figure 8.3 ▶ A particle with energy E penetrating a potential energy barrier.

There are three regions in which the particle may exist. Our first task is to determine the wave function for the particle in each region. The probability of finding the particle in a specific region is directly related to ψ^2 for the particle in that region. Because the wave function in region III is not zero, there is a finite probability that the particle can be in that region after tunneling through the barrier.

In treating the barrier penetration problem, the terminology used is very similar to that used in describing absorbance of light by a solution. We speak of the *intensity* of the incident particle–wave in region I and the intensity of the transmitted particle–wave in region III. In region II, there is *some* intensity of the penetrating particle–wave. In this case, we do not speak of the "particle" penetrating the barrier. It is only because of the particle–wave duality, the de Broglie wave character of a moving particle, that the penetration occurs. In the classical sense, the particle cannot exist within the barrier.

Although we will write the wave equation for the particle in each region, it will not be necessary to solve all of them in detail. It will be sufficient to write the solutions by comparing the equations to those we have already considered.

In region I, the particle moving toward the barrier can be described by a wave equation similar to that presented earlier for the particle in a box. Therefore, the wave equation can be written as

$$\frac{d^2\psi_{\mathrm{I}}}{dx^2} + \frac{2m}{\hbar^2}E\psi_{\mathrm{I}} = 0. \tag{8.5}$$

Replacing $2mE/\hbar^2$ by k^2, we can write the solution directly as

$$\psi_{\mathrm{I}} = A\mathrm{e}^{ikx}. \tag{8.6}$$

However, part of the particle–wave will be reflected by the barrier so we need to add a correction term to the wave function to account for this reflection. The form of that term will be similar to that already written except for the coefficient and the sign of the exponent. The final form of the wave function for the particle in region I is

$$\psi_{\mathrm{I}} = A\mathrm{e}^{ikx} + B\mathrm{e}^{-ikx}, \tag{8.7}$$

where A and B are the amplitudes of the incident and reflected waves, respectively. If A and B are real, B^2/A^2 gives the fraction of the particles (or fraction of the wave) that will be reflected by the barrier. If A and B are complex, B^*B/A^*A gives the fraction of the particles or wave reflected.

In region III, the wave equation can be written in the same form as that for the particle–wave in region I:

$$\frac{d^2\psi_{\mathrm{III}}}{dx^2} + \frac{2mE}{\hbar^2}\psi_{\mathrm{III}} = 0. \tag{8.8}$$

The solution for this equation can be written as

$$\psi_{\mathrm{III}} = J\mathrm{e}^{ikx}, \tag{8.9}$$

where J is the amplitude of the wave in region III.

For region II, the wave equation can be written as

$$\frac{d^2\psi_{\mathrm{II}}}{dx^2} + \frac{2mE}{\hbar^2}(U_0 - E)\,\psi_{\mathrm{II}} = 0. \tag{8.10}$$

There are two components to ψ_{II} since the wave penetrates the left-hand side of the barrier, but it can be partially reflected by the right-hand surface of the barrier. Therefore, in region II the wave function can be written as

$$\psi_{\mathrm{II}} = K\exp\left[\left(\frac{2m}{\hbar^2}(U_0 - E)\right)^{1/2} x\right] + L\exp\left[-\left(\frac{2m}{\hbar^2}(U_0 - E)\right)^{1/2} x\right], \tag{8.11}$$

which can be simplified by letting $j = [2m(U_0 - E)/\hbar^2]^{1/2}$,

$$\psi_{\mathrm{II}} = K\mathrm{e}^{ijx} + L\mathrm{e}^{-ijx}. \tag{8.12}$$

The various constants are determined from the behavior of the wave function at the boundaries. At the boundaries $x = 0$ and $x = a$, both ψ and $d\psi/dx$ are continuous (and have the same value) since the wave functions for the particle–wave in adjacent regions must join smoothly. Therefore

$$\psi_{\mathrm{I}}(0) = \psi_{\mathrm{II}}(0). \tag{8.13}$$

This condition leads to the result that

$$A + B = K + L. \tag{8.14}$$

Since

$$\frac{d\psi_{\mathrm{I}}(0)}{dx} = \frac{d\psi_{\mathrm{II}}(0)}{dx}, \tag{8.15}$$

it can be shown that

$$ikA - ikB = \sqrt{\frac{2m(U_0 - E)}{\hbar^2}}\,(K - L). \tag{8.16}$$

At the boundary a,

$$\psi_{\mathrm{II}}(a) = \psi_{\mathrm{III}}(a), \tag{8.17}$$

which allows us to show that

$$K \exp\left(\sqrt{\frac{2m(U_0 - E)}{\hbar^2}}a\right) + L \exp\left(-\sqrt{\frac{2m(U_0 - E)}{\hbar^2}}a\right) = Je^{ika}. \tag{8.18}$$

Finally, since

$$\frac{d\psi_{\mathrm{II}}(a)}{dx} = \frac{d\psi_{\mathrm{III}}(a)}{dx}, \tag{8.19}$$

it can be shown that

$$\sqrt{\frac{2m(U_0 - E)}{\hbar^2}}\left[\left(K \exp\sqrt{\frac{2m(U_0 - E)}{\hbar^2}}\right) - L \exp\left(-\sqrt{\frac{2m(U_0 - E)}{\hbar^2}}\right)\right] = ikJe^{ika}. \tag{8.20}$$

Although the detailed steps will not be shown here, it is possible from these relationships to evaluate the constants.

The *transparency* of the barrier is given by the probability density in region III (J^2) divided by that in region I (A^2),

$$T = \frac{J^2}{A^2} = \exp(-2a[2m(U_0 - E)/\hbar^2]^{1/2}). \tag{8.21}$$

If the barrier is not rectangular, the barrier height, U, must be expressed as a function of distance, x. Then the probability of barrier penetration is written in terms of the energy function, $U(x)$, as

$$T \simeq \exp\left(-\frac{2\sqrt{2m}}{\hbar}\int_0^a [U(x) - E]^{1/2}\, dx\right). \tag{8.22}$$

In this expression, the integral represents how much of the barrier lies above the energy level of the particle.

Several conclusions are reached immediately from the form of Eqs. (8.21) and (8.22). First, the transparency decreases as the thickness of the barrier, a, increases. Second, the transparency also decreases as the difference between the energy of the particle and the height of the barrier increases. That is, the transparency decreases when more of the barrier lies above the energy of the particle. When the energy of the particle is equal to the height of the barrier, the exponent becomes zero and the transparency is 1 (all of the particles can pass over the barrier in classical behavior). Finally, it should also be clear that the transparency of the barrier decreases with increasing mass, m, of the particle. It should also be pointed out that if $\hbar = 0$ (energy not quantized but rather classical in behavior), the exponent becomes 0 and there is no possibility of the particle getting past the barrier. Therefore, as stated earlier, tunneling is a quantum mechanical phenomenon. Tunneling is much more significant for light particles, and the later sections of this chapter show applications of the barrier penetration model to cases involving electron tunneling. It should now be clear why the walls of the one-dimensional box (see Chapter 3) had to be made infinitely high in order to confine the particle to the box 100% of the time.

8.3 Alpha Decay

Schrödinger's solution of the wave equation in 1926 opened the door to other applications of wave mechanics. For example, nuclear physicists had known that α decay occurs with the emitted α particle having an energy that is typically in the range 2–9 MeV (1 eV= 1.6×10^{-12} erg; 1 MeV= 10^6 eV). However, inside the nucleus, the α particle (a helium nucleus) is held in a potential energy well caused by its being bound to other nuclear particles. Further, in order to cause an α particle to penetrate the nucleus from the *outside*, the α particle would need to overcome the Coulomb repulsion. For a heavy nucleus ($Z = 80$), the Coulomb barrier will be about 25 MeV in height. Therefore, in order for an α particle *inside* the nucleus to escape, it would need to have an energy of at least that magnitude. However, for many cases, the α particles have energies of only 5–6 MeV. Figure 8.4 shows the energy relationships and parameters needed for a discussion of this problem. The question is, simply, how can the α particle be emitted with an energy of 5–6 MeV through a barrier that is as great as 25 MeV?

In 1928, only two years after Schrödinger's solution of the hydrogen atom problem, the problem of α decay was solved by Gurney and Condon and independently by G. Gamow. It was assumed that the α particle (two protons and two neutrons) moves inside the nucleus but is constrained by the potential barrier. Quantum mechanically, $\psi^*\psi$ predicts that there is *some* probability of finding the α particle on the outside of the barrier. Since the particle does not have sufficient energy to go over the barrier, it must escape by tunneling through the barrier.

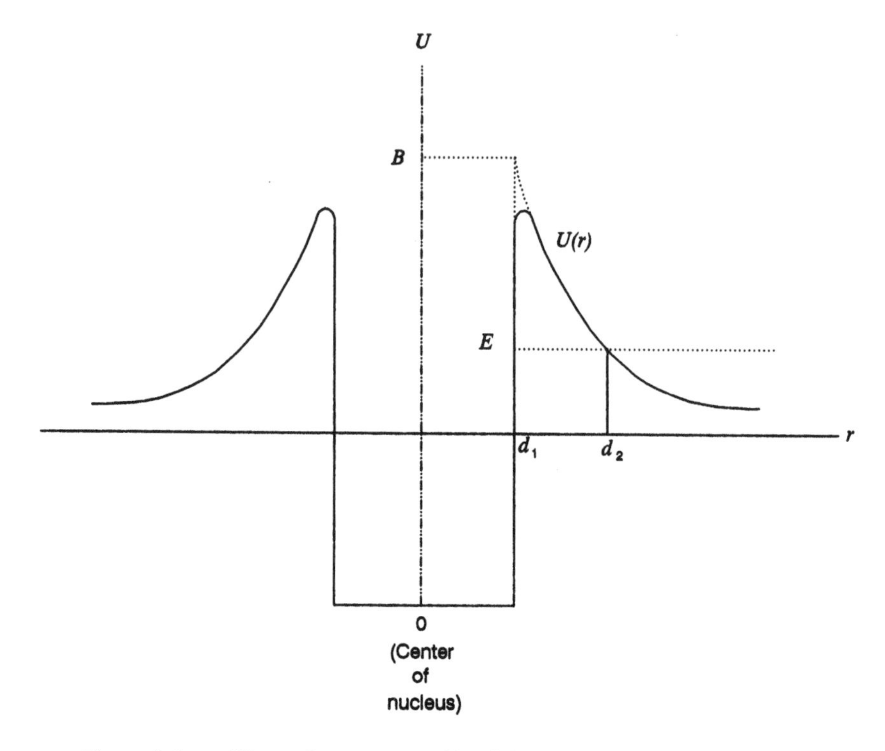

Figure 8.4 ▶ The nuclear energy well and Coulomb barrier for alpha decay.

The rate of decay of a nucleus can be expressed by the first-order equation

$$-\frac{dN}{dt} = kN, \tag{8.23}$$

where k is the *decay constant* and N is the number of nuclei. The rate of the decay process is reflected by the magnitude of k, which in turn is related to the transparency of the barrier. Therefore, the problem of explaining the observed decay constants for α decay can be solved if we can compute a transparency for the Coulomb barrier produced by the nucleus.

Treating the problem of a decay involves using an appropriate expression for $U(r)$ in Eq. (8.22). If the potential well inside the nucleus is assumed to have a square-bottom and the shape of the barrier is given by a Coulomb potential outside the nucleus [$U(r) = Zze^2/r$, where Z is the charge of the daughter nucleus and z is the charge on the α particle], the integral, I, becomes

$$I = \int_{d_1}^{d_2} (Zze^2 - Er)^{1/2} \frac{dr}{r^{1/2}}. \tag{8.24}$$

After a change of variables, $x = r^{1/2}$ and $q^2 = Zze^2/E$, we have $r = x^2$ and $dr = 2x\,dx$. Also, we can simplify the integral by realizing that

$$(Zze^2 - Er)^{1/2} = \left(\frac{Zze^2 E}{E} - Er\right)^{1/2} = E^{1/2}\left(\frac{Zze^2}{E} - r\right)^{1/2}. \quad (8.25)$$

Now Eq. (8.24) can be written as

$$I = 2\sqrt{E}\int_{d_1}^{d_2} \left(q^2 - x^2\right)^{1/2} dx. \quad (8.26)$$

This integral is of a form that can be found in tables of integrals,

$$\int \left(a^2 - x^2\right)^{1/2} dx = \frac{1}{2}\left[x\left(a^2 - x^2\right)^{1/2} + a^2 \sin^{-1}\left(\frac{x}{a}\right)\right]. \quad (8.27)$$

Therefore, the integral shown in Eq. (8.26) becomes

$$I = \sqrt{E}\left[x\left(q^2 - x^2\right)^{1/2} + q^2 \sin^{-1}\left(\frac{x}{q}\right)\right]_{d_1}^{d_2}. \quad (8.28)$$

The values for d_1 and d_2 are related to the charges and energies by Coulomb's law,

$$E = \frac{Zze^2}{d_2} \quad \text{and} \quad B = \frac{Zze^2}{d_1}. \quad (8.29)$$

We need not show the details of the derivation here, but it is possible to derive an expression for the decay constant, k, which is

$$k = \frac{\hbar}{2md_1^2}\exp\left[-\frac{8\pi Zze^2}{hv}\left(\cos^{-1}\left(\frac{E}{B}\right)^{1/2} - \left(\frac{E}{B}\right)^{1/2}\left(1 - \frac{E}{B}\right)^{1/2}\right)\right], \quad (8.30)$$

where the symbols have already been explained except for v, the velocity of the α particle in the nucleus. This quantity comes into consideration because the number of times the particle moves back and forth in the nucleus and comes in contact with the barrier is related to the probability that it will eventually penetrate the barrier. Using appropriate expressions for the values of the parameters, the calculated decay constants are generally in excellent agreement with those observed experimentally. For example, the calculated and experimental values of k (in s^{-1}) for a few emitters of α particles are as follows: for ^{148}Gd, $k_{calc} = 2.6 \times 10^{-10}$, $k_{exp} = 2.2 \times 10^{-10}$; for ^{214}Po, $k_{calc} = 4.9 \times 10^3$, $k_{exp} = 4.23 \times 10^3$; and for ^{230}Th, $k_{calc} = 1.7 \times 10^{-13}$, $k_{exp} = 2.09 \times 10^{-13}$. The application of the barrier penetration model to α decay has been quite successful.

8.4 Tunneling and Superconductivity

While the model of tunneling of particles through barriers has been successfully applied to several phenomena, it is perhaps in the area of superconductivity that tunneling is most important. Certainly this is so in regard to technology, and we will briefly describe how this application of barrier penetration is so important. The discussion here will include only the rudiments of this important and timely topic. For a more complete discussion of superconductivity, consult the references at the end of this chapter, especially the works by Kittel and Serway.

If two metal strips are separated by an insulator, no electric current passes through the system in normal circumstances. The insulator acts as a barrier to the particles (electrons) in a manner analogous to that discussed in Sections 8.1 and 8.2. Often the barrier is an oxide layer on one of the metals. If the insulator is made sufficiently thin (1–2 nm), it is possible for electrons to tunnel through the barrier from one metal to the other. For metals not behaving as superconductors, the conductivity through the barrier follows Ohm's law, which indicates that the current is directly proportional to the voltage. This type of tunneling by electrons is known as *single-particle tunneling*.

Heike Kamerlingh Onnes, a Dutch physicist, liquified helium (bp 4.2 K) in 1908. This was an important event since the first superconductors studied did not become "super" conducting except at very low temperatures. Superconductivity was discovered in 1911 by Onnes. After studying the resistivity of platinum, mercury was studied because it could be obtained in very high purity. It was found that at 4.15 K (the temperature at which mercury becomes superconducting, T_C) the resistivity of mercury dropped to 0. In 1933, Walther Meissner and R. Ochsenfeld found that when certain types of superconductors were kept below their critical temperatures in a magnetic field, the magnetic flux was expelled from the interior of the superconductor. This behavior is known as the *Meissner effect*.

In 1957, J. Bardeen, L. N. Cooper, and J. R. Schrieffer developed a theory of superconductivity known as the *BCS theory*. According to this theory, electrons are coupled to give pairs having a resultant angular momentum of 0. These electron pairs are known as *Cooper pairs*, and their characteristics are responsible for some of the important properties of superconductors. In 1962, B. Josephson predicted that two superconductors separated by a barrier consisting of a thin insulator should be able to have an electric current between them due to tunneling of Cooper pairs. This phenomenon is known as the *Josephson effect* or *Josephson tunneling*. The existence of superconductors that have T_C values higher than the boiling point of liquid nitrogen is enormously important since liquid helium is not required for cooling them.

The migration of electrons through a solid is impeded by motion of the lattice members as they vibrate about their equilibrium positions. In the case of a metal, the lattice sites are occupied by metal ions. In Chapter 6, we saw that the frequency of vibrations is given by

$$v = \frac{1}{2\pi}\sqrt{\frac{k}{m}}, \tag{8.31}$$

where k is the *force constant* or *spring constant* and m is the mass. Assuming that passage of electrons through the metal is impeded by the vibrational motion of the metal ions, we would assume that for a given metal, the T_C would vary with $1/m^{1/2}$ for the metal, as it does for classical conductivity. The lighter the atom, the greater the vibrational frequency and the more the motion of the electron would be hindered. Experimental measurements of the T_C have been carried out for ^{199}Hg, ^{200}Hg, and ^{204}Hg, for which the T_C values are 4.161, 4.153, and 4.126 K, respectively. Figure 8.5 shows that a graph of T_C versus $m^{1/2}$ is linear, which is the inverse of the expected relationship.

If we suppose a metal atom to be about 10^{-8} cm in diameter and that the electron is moving with a linear velocity of about 10^8 cm/s, the electron will be in the proximity of the metal atom for about $(10^{-8}$ cm$)/(10^8$ cm/s$)$,

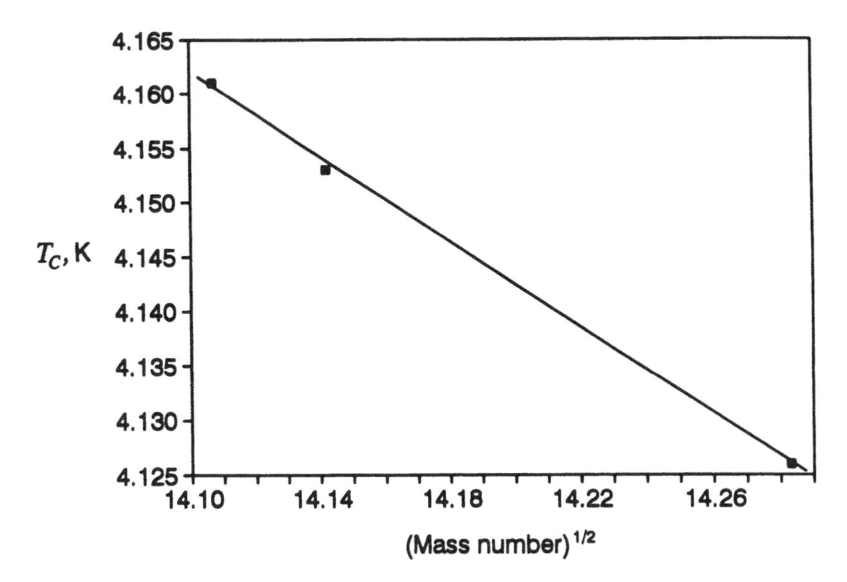

Figure 8.5 ▶ The relationship between the T_C and isotopic mass for isotopes of mercury.

or a time on the order of 10^{-16} s. For many solids, the vibrational frequency is on the order of 10^{12} s^{-1} or the oscillation time is on the order of 10^{-12} s. Therefore, the lattice reorientation (mass number) time is very long compared to the time it takes for the electron to pass this site in the lattice.

A very elementary view of a metal consists of metal ions at lattice sites with mobile electrons moving through the solid in conduction bands. The motion of Cooper pairs through a metal lattice is thought to be linked to the lattice motion just described. One view is that as one electron passes between two metal ions, the ions move inward from their respective lattice sites. This region thus has an instantaneous increase in positive charge. This analogy is very similar to that of instantaneous dipoles in helium atoms when the two electrons are found at some instant on the same side of the atom, giving rise to *London dispersion forces*. The increased positive region exerts an attractive force on a second electron, which follows the first through the opening between the metal sites before lattice reorganization occurs. The effect is that two electrons behave as a pair (the Cooper pair) having opposing spins but existing as an entity, having a resultant spin angular momentum of 0.

Tunneling in superconductors involves the behavior of Cooper pairs, which, because of their resulting zero spin, behave as bosons. Thus, they are not required to obey the Pauli exclusion principle as fermions do. Therefore, any number of Cooper pairs can populate the same state. The BCS theory incorporates the idea that all of the electrons form a ground state consisting of Cooper pairs. In the conductivity of normal metals, the lattice vibration of the atoms reduces the mobility of the electrons, thereby reducing conductivity. In the case of Cooper pairs, reducing the momentum of one pair requires the reduction of momentum for all the pairs in the ground state. Since this does not occur, lattice vibrations do not reduce the conductivity that occurs by motion of Cooper pairs. For conduction in normal metals, lattice vibrations decrease conductivity. For superconductivity, the lattice motion is responsible for the formation of Cooper pairs, which gives rise to superconductivity.

As a part of his description of Cooper pairs, Josephson predicted that two superconductors separated by a thin insulating barrier could experience pair tunneling. One result of this phenomenon would be that the pairs could tunnel without resistance, yielding a direct current with no applied electric or magnetic field. This is known as the *dc Josephson effect*.

Figure 8.6 shows the arrangement known as a Josephson junction that leads to the dc Josephson effect. Cooper pairs in one of the superconductors are described by the wave function ψ_1, and those in the other are

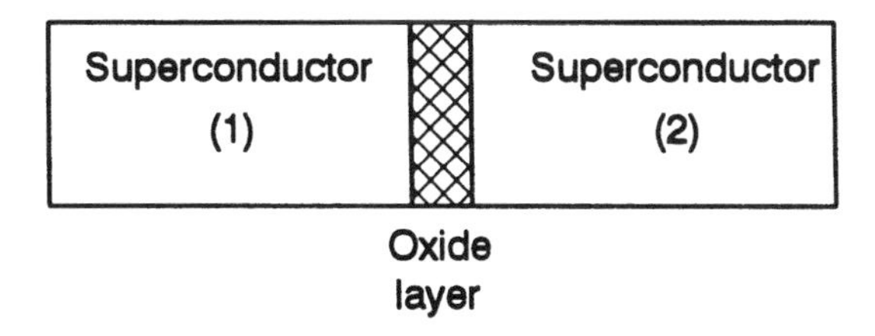

Figure 8.6 ▶ A schematic diagram of a Josephson junction.

described by the wave function ψ_2. The appropriate Hamiltonian can be written as

$$-\frac{\hbar}{i}\frac{\partial}{\partial t}\psi = \hat{H}\psi, \tag{8.32}$$

so that when both superconductors are considered,

$$-\frac{\hbar}{i}\frac{\partial \psi_1}{\partial t} = \hbar T \psi_2 \quad \text{and} \quad -\frac{\hbar}{i}\frac{\partial \psi_2}{\partial t} = \hbar T \psi_1, \tag{8.33}$$

where T is the rate of current flow across the junction from each of the superconductors. If the insulator is too thick for tunneling to occur, $T = 0$. It is possible to show that

$$I = I_{\text{m}} \sin(\phi_2 - \phi_1) = I_{\text{m}} \sin \delta, \tag{8.34}$$

where I_{m} is the maximum current across the junction when there is no applied voltage. In Eq. (8.34), ϕ is the phase of the pair, and all pairs in a given superconductor have the same phase. When the applied voltage is zero, the current varies from I_{m} to $-I_{\text{m}}$, depending on the phase difference for the two superconductors. The dc Josephson effect is one of the results of pair tunneling.

If a dc voltage is applied across the junction between two superconductors, an alternating current oscillates across the junction. It can be shown (e.g., see Kittel, 1996) that the oscillating current, I, can be expressed by

$$I = I_{\text{m}} \sin\left(\phi\,(0) - \frac{2eVt}{\hbar}\right), \tag{8.35}$$

where $\phi\,(0)$ is a constant, the phase at time zero, V is the voltage, t is the time, and e is the electron charge. This phenomenon is referred to as *ac Josephson tunneling*.

When dc tunneling occurs in the presence of an external magnetic field, a periodic tunneling process occurs. Linking two of the Josephson junctions together in parallel allows an interference effect to be observed that is very sensitive to the magnetic field experienced by the system. Such a system is known as a *s*uperconductivity *qu*antum *i*nterference *d*evice (SQUID) and such devices are used to detect very weak magnetic fields. For instance, they have been used to study the fields produced by neuron currents in the human brain.

Tunneling by Cooper pairs plays an important role in the behavior of superconductors. As superconductors having higher and higher T_C values are obtained, it is likely that this tunneling behavior will be exploited in technological advances.

8.5 The Scanning Tunneling Microscope

An important application of tunneling involves the *scanning tunneling microscope* (STM) invented by G. Binnig and H. Rohrer in 1981. In this case, the tunneling is by electrons that tunnel between the surface of a solid and the tip of a probe maintained at a very short distance from the surface. A typical gap of a few angstroms is maintained by a feedback loop that maintains a constant current flow between the solid surface and the tip of the movable probe by adjusting the distance between them (the height of the barrier). A simple schematic diagram of a scanning tunneling microscope is shown in Figure 8.7. The tunneling occurs because the wave functions for electrons in the solid do not end abruptly at the surface of the solid, but rather extend into the space above the surface of the solid. This behavior

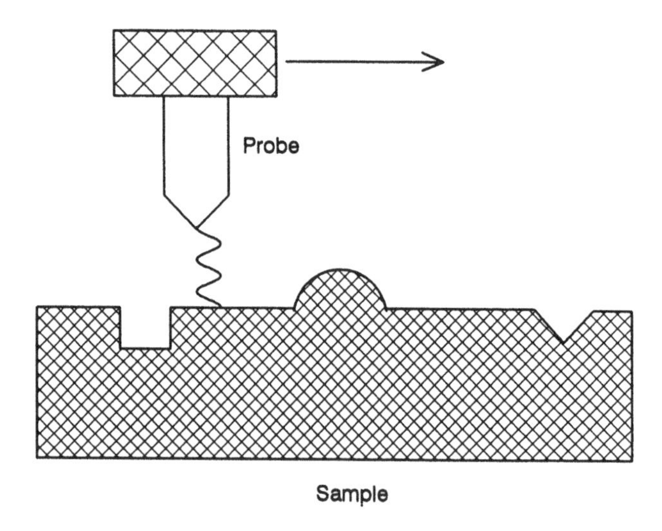

Figure 8.7 ▶ Schematic diagram of a scanning tunneling microscope.

is similar to that of the harmonic oscillator in which the wave functions do not end at the classical boundaries of the oscillation. Moreover, a particle in a one-dimensional box would not give wave functions that terminate at the walls of the box if the walls were not infinitely high.

As the tip of the probe moves across the surface of the solid, its vertical motion gives a plot of the surface features in the form of a contour map. The resolutions of features are on the order of 2 Å, but surface height differences of about 0.01 Å can be detected. Because the electron wave functions extend into space above the solid, the STM can literally map the wave functions of surface atoms and "see" individual atoms. Therefore, the STM can locate the active sites on catalyst surfaces, and it has enabled surfaces of materials ranging from metals to viruses to be mapped.

8.6 Spin Tunneling

Perhaps the most recent observation of tunneling involves that of the tunneling of electron spins through the potential barrier that separates one spin orientation from the other. In this case, a solid complex compound of manganese having the complete formula $[Mn_{12}(CH_3COO)_{16}(H_2O)_4O_{10}] \cdot 2CH_3COOH \cdot 4H_2O$ was studied in a magnetic field at low temperature. The molecule has a spin state of 10 that is spread over the 12 manganese ions. In the absence of a magnetic field, there are two sets of energy levels that involve the orientations from $+10$ to -10. There are 21 different orientations, $0, \pm 1, \ldots, \pm 10$, and the crystal of the compound is not isotropic in its magnetization. Thus, the energies associated with the orientations of the spin vectors have different values and constitute two identical sets of levels as shown in Figure 8.8. When the crystal is placed in a magnetic field that is applied in the positive direction (parallel to the + crystal axis direction), the states in the parallel direction are decreased in energy, while those in the opposite direction are increased in energy (see Figure 8.8).

Tunneling of spins between the two sets of levels occurs at very low temperatures ($2 - 3$ K) as *resonant tunneling* under the conditions where the states are at the same energy (no magnetic field applied). If the magnetic field is applied, it is possible to cause the energy levels to change so that some of the states are again at the same energy (a sort of accidental degeneracy but one that depends on an external field). The states that are at the same energy under these conditions do not have the same numerical value of the spin quantum number. For example, the $+3$ state may reside at the same energy as the -4 state, as shown in Figure 8.8.

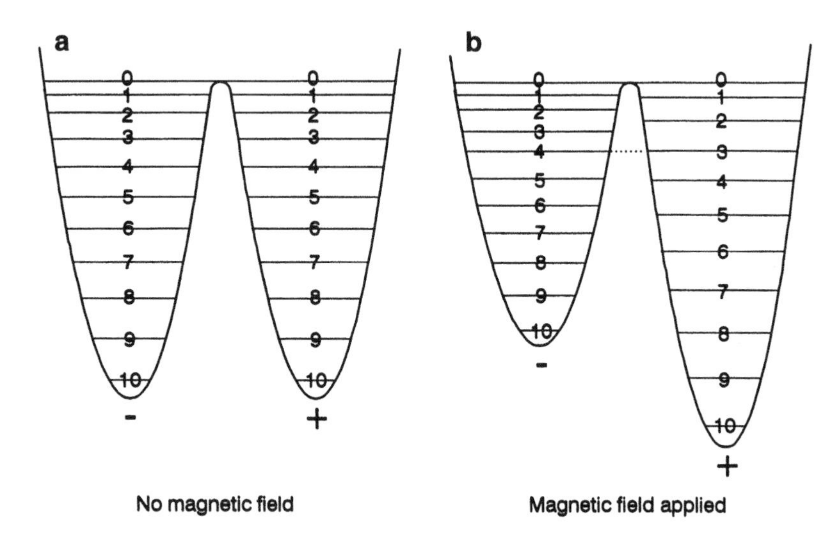

Figure 8.8 ▶ Spin states in the solid manganese–acetate complex.

If all the spins are forced to the -10 state by applying an external field, decreasing the field will allow resonant tunneling at some specific magnitude of the applied field as some of the levels are again brought to the same energy. Therefore, when the property of magnetization is studied at various values of the applied external field, the hysteresis loop, unlike the classical plot of magnetization versus applied field, shows a series of steps. These steps are observed because quantum mechanical tunneling is occurring. A temperature effect of this phenomenon indicates that the spins must be populating states of about $m = \pm 3$ by thermal energy. Otherwise, the tunneling times from a potential well as deep as those of the $m = \pm 10$ states would be extremely slow with transition probabilities that are very low. The steps in the hysteresis plot clearly show that quantum mechanical tunneling occurs as electrons go from one spin state to another without passing over the barrier between them. At higher temperatures, the electrons can pass over the barrier without tunneling being necessary.

8.7 Tunneling in Ammonia Inversion

The pyramidal ammonia molecule has associated with it a vibration in which the molecule is turned inside out. This vibration, known as *inversion*, is shown in Figure 8.9 and has a frequency on the order of 10^{10} s^{-1}. Excitation of the vibration from the first to the second vibrational level gives

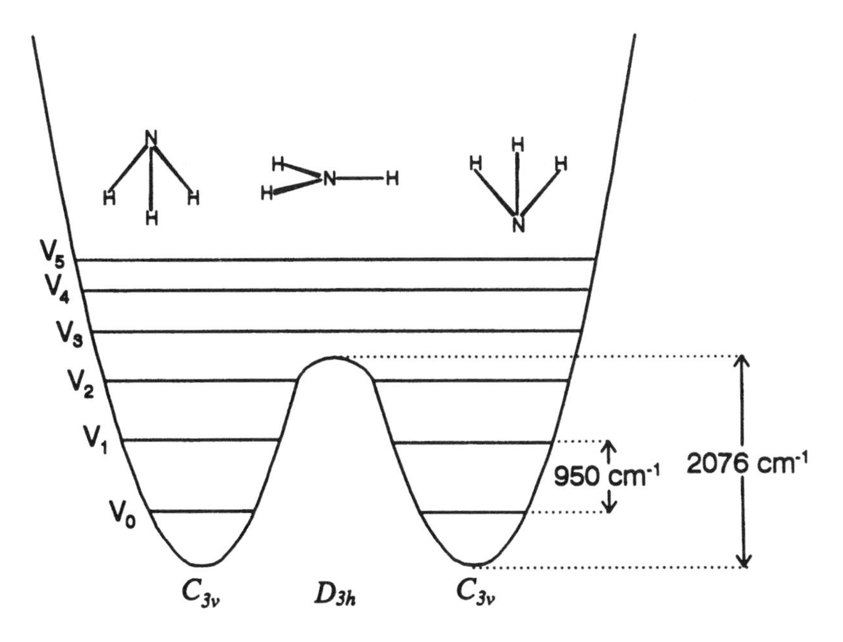

Figure 8.9 ▶ The inversion of the ammonia molecule.

rise to an absorption at 950 cm^{-1}, and the barrier height is 2076 cm^{-1}. According to the Boltzmann distribution law, the population of the molecules in the lowest two levels (n_0 and n_1) should be given by

$$\frac{n_1}{n_0} = \mathrm{e}^{-\Delta E/RT}, \tag{8.36}$$

where ΔE is the difference in energy between the two states, T is the temperature (K), and R is the molar gas constant. In this case,

$$\begin{aligned} E &= hc\bar{\nu} \\ &= \left(6.63 \times 10^{-27} \text{ erg s}\right) \times \left(3.00 \times 10^{10} \text{ cm/s}\right) \times \left(950 \text{ cm}^{-1}\right) \\ &= 1.89 \times 10^{-13} \text{ erg.} \end{aligned}$$

This energy per molecule can be converted to J/mol by multiplying by Avogadro's number and dividing by 10^7 erg/J. In this case, the energy is 11,400 J/mol. Therefore, at room temperature (300 K),

$$\frac{n_1}{n_0} = \mathrm{e}^{(-11{,}400 \text{ J/mol})/[(8.3144 \text{ J/mol K}) \times (300 \text{ K})]} = 0.0105. \tag{8.37}$$

Consequently, almost all of the molecules will be in the lowest vibrational level. Since the barrier height is 2076 cm^{-1}, the molecules must invert by tunneling through the relatively low and "thin" barrier that will be transparent.

The de Broglie hypothesis was applied to the hydrogen atom by Schrödinger in 1926, and the wave character of moving electrons was demonstrated experimentally in 1927. Very soon thereafter, various applications of quantum mechanics started to be explored, and tunneling was used as a model for alpha decay by Gurney and Condon and also by Gamow in 1928. Consequently, tunneling has been a viable model for quantum phenomena almost since the beginning of quantum mechanics. The latest demonstration of this type of behavior appears to be the tunneling between spin states first described in 1996. From these and other applications, it should be apparent that tunneling needs to be discussed along with the harmonic oscillator, rigid rotor, particle in a box, and other topics in the study of quantum mechanical models.

References for further Reading

- Fermi, E. (1950). *Nuclear Physics*. Univ. of Chicago Press, Chicago. This book of lecture notes is based on courses taught by Fermi at the University of Chicago in 1949. Now available as an inexpensive paperback edition from Midway Reprints.
- Friedlander, G., Kennedy, J. W., Macias, E. S., and Miller, J. M. (1981). *Nuclear and Radiochemistry*, 3rd ed. Wiley, New York. A monumental book on nuclear chemistry with a thorough discussion of α decay.
- Kittel, C. (1996). *Introduction to Solid State Physics*, 7th ed., Chap. 12. Wiley, New York. The standard text on solid state physics with good coverage of the theories of superconductivity and Josephson tunneling.
- Serway, R. E. (2000). *Physics for Scientists and Engineers*, 5th ed. Saunders, Philadelphia. A thorough introduction to superconductivity and tunneling phenomena that is also very well written.
- Swarzchild, B. (1997). *Phys. Today* **50**, 17. A summary of spin tunneling work with references to the original literature.
- Tinkham, M. (1975). *Introduction to Superconductivity*. McGraw-Hill, New York. A fundamental book on superconductivity.

Problems

1. Suppose a rectangular potential barrier of height 2.0 eV and thickness 10^{-8} cm has an electron approach it. If the electron has an energy of 0.25 eV, what is the transmission coefficient? If the electron has an energy of 0.50 or 0.75 eV, what are the transmission coefficients?

2. Repeat Problem 1 for a neutron approaching the barrier.

3. Repeat Problem 1 for a helium atom approaching the barrier.

4. If an electron having a kinetic energy of 25 eV approaches a rectangular barrier that is 10^{-8} cm thick and has a height of 35 eV, what is the probability that the electron will penetrate the barrier? What will be the probability of the electron penetrating the barrier if the thickness is 10^{-7} cm?

5. Repeat the calculations in Problem 4 if a proton having an energy of 25 eV approaches the barrier.

6. Suppose that an electron having an energy of 10 eV approaches a rectangular barrier of 10^{-8} cm thickness. If the transmission coefficient is 0.050, how high is the barrier?

7. Consider a potential barrier represented as follows:

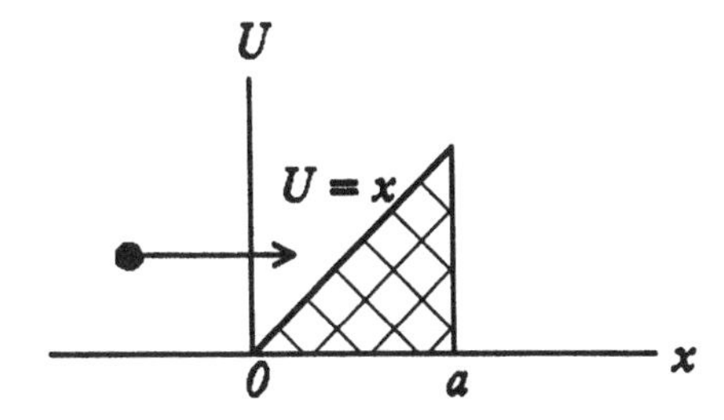

Determine the transmission coefficient as a function of particle energy.

8. An electron having an energy of 15 eV impinges on a barrier 25 eV in height.

(a) Determine the probability of the electron tunneling through the barrier if the barrier is 1.20 nm in thickness.

(b) What will be the probability that the electron will tunnel through the barrier if its thickness is 0.120 nm?

9. Suppose a proton is bound in a nucleus in a potential that is approximately a square well with walls that are infinitely high. Calculate the wave length and energy for the emitted photon when the proton falls from $n = 3$ to the $n = 2$ state if the nucleus has a diameter of 1.50×10^{-13} cm.

Chapter 9

Diatomic Molecules

In describing the characteristics of matter by means of quantum mechanics, we have progressed from the behavior of particles in boxes to atoms. The next step in the development is to consider molecules, and the last four chapters of this book deal with descriptions of molecules. As we shall see, there are two major approaches to the problem of molecular quantum mechanics. These are the valence bond approach developed by Heitler, London, Slater, and Pauling, and the molecular orbital (MO) approach developed primarily by Robert Mulliken. Most of what we will cover in this chapter deals with the molecular orbital approach because it is much simpler to use from a computational standpoint.

9.1 An Elementary Look at a Covalent Bond

If we begin the task of describing how molecules exist, it is useful to look first at a very simple picture. Consider the H_2 molecule to be forming as we bring two hydrogen atoms together from a very large distance as shown in Figure 9.1. As the atoms get closer together, the interaction between them increases until a covalent bond forms. Each atom is considered complete in itself, but an additional interaction occurs until we can represent the molecule as shown in Figure 9.1(c). What do we know about this system? First, the interaction between a hydrogen nucleus and its electron is -13.6 eV, the binding energy of an electron in a hydrogen atom. The hydrogen *molecule* has a bond energy of 4.51 eV (required to break the bond or -4.51 eV when the bond forms). If we represent the H_2 molecule as

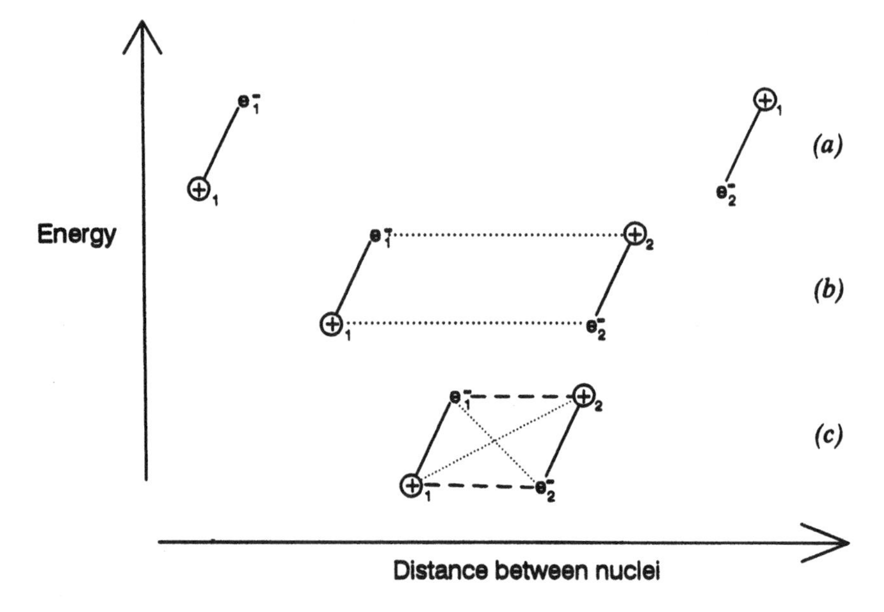

Figure 9.1 ▶ The formation of a chemical bond as atoms approach.

shown in Figure 9.1(c), it becomes clear that the molecule differs from two atoms by the two dashed lines in the figure and by the repulsion of the two nuclei and the two electrons. The dashed lines represent the interactions of electron (1) with nucleus (2) and electron (2) with nucleus (1). Despite the repulsions between the nuclei and between the electrons, the bond energy for the hydrogen molecule is 4.51 eV/bond (432 kJ/mol). Therefore, the attractions between the nuclei and the electrons more than offset the repulsions at the normal bond length of 0.76 Å (76 pm).

After the quantum mechanical treatment of the models shown in earlier chapters, we do not expect that the quantum mechanical treatment of even diatomic molecules will be simple. The preceding discussion was intended to show that we intuitively know a great deal about diatomic molecules. Our reference points are provided by the energies of electrons in atoms and the bond energies of the molecules.

The energies of chemical bonds are usually expressed as positive numbers representing the energy required to break the bond:

$$A : B \longrightarrow A + B, \quad \Delta H = \text{bond enthalpy (positive)}. \tag{9.1}$$

When the bond forms, the same quantity of energy is involved, but it is liberated (this is the negative of the bond enthalpy).

9.2 Some Simple Relationships for Bonds

Let us begin by considering several molecules that are relatively simple from the standpoint of the orbitals used. Because *s* orbitals are singly degenerate and spherically symmetric, the simplest bonds are those where only *s* orbitals are used. Fortunately, we have a range of molecules to consider. These include H_2, Li_2 ... Cs_2, LiH ... CsH, NaLi, KNa, and RbNa, all bound with single bonds. Searching for relationships between properties of molecules has long been an honorable activity for persons seeking to understand the chemical bond, and the chemical literature contains an enormous number of such relationships. Moreover, many of these relationships (whether they be empirical, semiempirical, or theoretical) are of great heuristic value. Before engaging the gears of the quantum machinery, we will apply a little chemical intuition.

Our intuition tells us that if everything else is equal, the longer a bond is, the *weaker* it is likely to be. We expect this because even the most naive view of orbital interaction (overlap) would lead us to the conclusion that larger, more diffuse orbitals do not interact as well as smaller, more compact ones. For our first attempt at verifying our intuition, we will make a graph of bond energy versus bond length for the molecules previously listed. Table 9.1 shows the pertinent data for the atoms considered. The graph that results is shown in Figure 9.2. Although we have simplified our

TABLE 9.1 ► Properties of Molecules Bonded by *s* Orbitals

Molecule	Bond length (pm)	Bond energy (kJ/mol)	Average IP[a] (kJ/mol)
H_2	74	432	1312
Li_2	267	105	520
Na_2	308	72.4	496
K_2	392	49.4	419
Rb_2	422	45.2	403
Cs_2	450	43.5	376
LiH	160	234	826
NaH	189	202	807
KH	224	180	741
RbH	237	163	727
CsH	249	176	702
NaLi	281	88.1	508
NaK	347	63.6	456
NaRb	359	60.2	447

[a] The ionization potentials (IP) of H, Li, Na, K, Rb, and Cs atoms are simply the average values for the diatomic molecules.

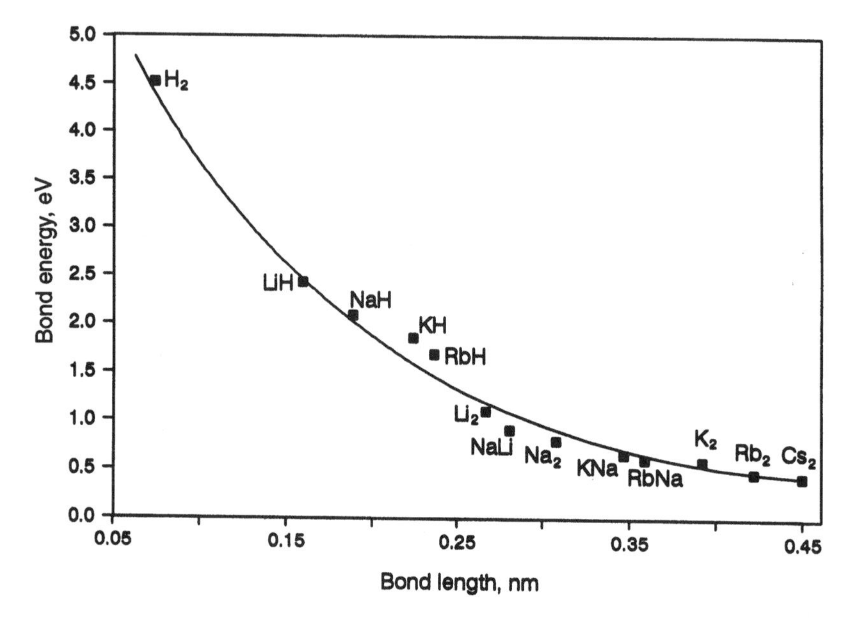

Figure 9.2 ▶ The relationship between bond length and bond energy for molecules bonded by overlap of *s* orbitals.

study by considering only those molecules using *s* orbitals in bonding, it is obvious that there is, as expected, a reasonably good relationship between bond energy and bond length for such cases. A relationship like that shown in Figure 9.2 confirms that some of our elementary ideas about chemical bonds are correct. We could now find the equation for the line to obtain an empirical algebraic relationship if we wished.

Having "discovered" one relationship that fits bonds between atoms using *s* orbitals, we will now take a different approach. We know that metals, particularly those in Group IA, have low ionization potentials and therefore do not have great attraction for electrons. If two alkali metals form a bond, we would expect the bond to be weak. However, we will not limit our discussion to alkali metals but rather include all the molecules for which data are shown in Table 9.1.

As a first and very crude approximation, we can consider an electron pair bond between atoms A and B as a mutual attraction of these atoms for the pair of electrons. We know that the attraction of an atom A for its outer electron is measured by its ionization potential. The same situation exists for atom B. Since the strength of the bond between atoms A and B reflects their mutual attraction for the electron pair, we might guess that the electron pair would be attracted by the two atoms with an energy that is related to the average of the ionization potentials of atoms A and B. Life is not usually this simple so we hope that the bond energy will be *related* to the average ionization in some simple way.

The first problem we face is how to calculate the average ionization potential of two atoms. The two approaches to getting a mean value for two parameters are the *arithmetic mean*, $(x_1 + x_2)/2$, and the *geometric mean*, $(x_1 x_2)^{1/2}$. When two quantities are of similar magnitudes, the two averages are about the same. In fact, if $x_1 = x_2$, the averages are identical. On the other hand, if one of the quantities is zero ($x_2 = 0$), the arithmetic mean is $x_1/2$, but the geometric mean is 0. Because bonds between atoms having greatly differing ionization potentials are to be considered, we choose to use the geometric mean of the ionization potentials.

Figure 9.3 shows a graph of the average ionization potential versus bond energy for a series of diatomic molecules where only s orbitals are used in bonding. In this case, the relationship is indeed very good. Linear regression applied to the data gives the equation

$$\text{IP}_{\text{av}}\ (\text{kJ/mol}) = 2.352E\ (\text{kJ/mol}) + 302.2 \quad (r = 0.996). \tag{9.2}$$

Not only is our intuition correct, we have succeeded in obtaining a relationship that can be used for predictive purposes. For example, francium is a radioactive element for which relatively little data exists. The ionization

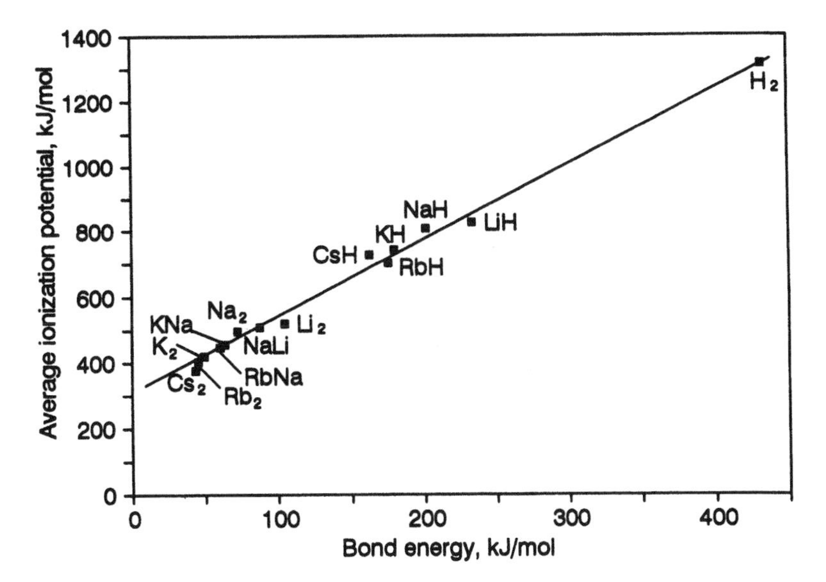

Figure 9.3 ▶ The relationship between the average ionization potential (geometric mean) and the bond energy of molecules having overlap of s orbitals.

potential for francium is about 3.83 eV or 369 kJ/mol. Using our relationship shown in Eq. (9.2), we predict a bond energy of only 28.5 kJ/mol (6.81 kcal/mol or 0.295 eV/bond) for the Fr_2 molecule. In a similar manner, we could estimate the bond energies of CsK, CsLi, etc.

It is interesting to speculate on what the intercept of 302.2 kJ/mol means. One interpretation is that the intercept, which corresponds to a hypothetical bond energy of 0, occurs when two atoms have such a low average ionization potential (302.2 kJ/mol) that they have no residual attraction for other electrons. The atom having the lowest ionization potential is Fr (369 kJ/mol) and the Fr_2 bond is very weak. If there were two atoms having an average ionization energy of 302.2 kJ/mol, their attraction for a bonding pair of electrons would be so slight that they should form no bond between them. Of course this assumes that the relationship is valid outside the range for which data are available to test it.

In this instance, our intuition that the ionization potentials of the atoms forming the bonds ought to be related to the bond energy is completely justified. Of course we restricted our examination to molecules using only *s* orbitals, but the results are still gratifying. Any relationship between bond energies and atomic properties that gives a correlation coefficient of 0.996 is interesting. Throughout the study of the chemical bond, many workers have sought to correlate bond energy with such properties as difference in electronegativity between the atoms and force constants for bond stretching. In most cases, a rather restricted list of molecules must be considered if a reasonable relationship is expected. We will now turn our attention to a quantum mechanical description of diatomic molecules.

9.3 The LCAO–MO Method

The linear combination of atomic orbitals–molecular orbital (LCAO–MO) method is based on the idea that a wave function for a molecule can be written as a *linear combination* of atomic wave functions. This can be expressed as

$$\psi = \sum a_i \phi_i. \tag{9.3}$$

For a diatomic molecule, this reduces to

$$\psi = a_1\phi_1 + a_2\phi_2, \tag{9.4}$$

where ψ is an atomic wave function and a is a weighting or mixing coefficient. As we shall see, $a_1\phi_1 - a_2\phi_2$ is also a possible linear combination. The values of the coefficients must be determined, and they are treated as parameters to be obtained by optimization using the variation method (see Section 4.4, p. 70). In the variation method, we begin by representing the energy using

$$E = \frac{\int \psi^* \hat{H} \psi \, d\tau}{\int \psi^* \psi \, d\tau}. \tag{9.5}$$

Substituting the trial wave function for ψ gives

$$E = \frac{\int \left(a_1\phi_1^* + a_2\phi_2^*\right) \hat{H} \left(a_1\phi_1 + a_2\phi_2\right) \, d\tau}{\int \left(a_1\phi_1^* + a_2\phi_2^*\right) \left(a_1\phi_1 + a_2\phi_2\right) \, d\tau}. \tag{9.6}$$

Expansion of the binomials leads to

$$E = \frac{a_1^2 \int \phi_1^* \hat{H} \phi_1 \, d\tau + 2a_1a_2 \int \phi_1^* \hat{H} \phi_2 \, d\tau + a_2^2 \int \phi_2^* \hat{H} \phi_2 \, d\tau}{a_1^2 \int \phi_1^* \phi_1 \, d\tau + 2a_1a_2 \int \phi_1^* \phi_2 d\tau + a_2^2 \int \phi_2^* \phi_2 \, d\tau}. \tag{9.7}$$

We have already assumed that

$$\int \phi_1^* \hat{H} \phi_2 \, d\tau = \int \phi_2^* \hat{H} \phi_1 \, d\tau \tag{9.8}$$

and that

$$\int \phi_1^* \phi_2 \, d\tau = \int \phi_2^* \phi_1 \, d\tau. \tag{9.9}$$

In other words, we have restricted the development to a *homonuclear* diatomic molecule. By inspection of Eq. (9.8) we see that we are going to have to deal with integrals like

$$\int \phi_1^* \hat{H} \phi_1 \, d\tau = H_{11} \tag{9.10}$$

$$\int \phi_1^* \hat{H} \phi_2 \, d\tau = H_{12}, \text{ etc.} \tag{9.11}$$

These integrals are frequently represented as $\langle \phi_1^* \mid \hat{H} \mid \phi_1 \rangle$ and $\langle \phi_1^* \mid \hat{H} \mid \phi_2 \rangle$, respectively, in a kind of shorthand notation. Similarly, the expansion of Eq. (9.7) leads to integrals of the type

$$\int \phi_1^* \phi_1 \, d\tau = S_{11} \tag{9.12}$$

$$\int \phi_1^* \phi_2 \, d\tau = S_{12}, \text{ etc.,} \tag{9.13}$$

which are written in shorthand notation as $\langle \phi_1^* \mid \phi_1 \rangle$ and $\langle \phi_1^* \mid \phi_2 \rangle$, respectively. We need to examine now what these types of integrals mean.

First, the integrals written as H_{11} and H_{22} represent the energies with which an electron is held in atoms 1 and 2, respectively. We know this because these integrals contain the wave function for an electron in each atom and the Hamiltonian operator is the operator for total energy. Moreover, the binding energy of an electron in an atom is simply the reverse of its ionization potential, which is *positive* since it requires work to remove an electron from an atom. Therefore, these binding energies are *negative*, with their magnitudes being determined from the *valence state ionization potential* (VSIP), an application of *Koopmans' theorem*, which states that ionization potential is equal in magnitude to the orbital energy. These integrals represent the Coulombic attraction for an electron in an atom and are accordingly called *Coulomb integrals*.

In a very loose way, the integrals represented as H_{12} and H_{21} are indicative of the attraction that nucleus 1 has for electron 2 and vice versa. These integrals are called the *exchange integrals* since they represent types of "exchange" attractions between the two atoms. We will see that these integrals are of paramount importance in determining the bond energy for a molecule. Determining the magnitude of the exchange integrals represents a good part of the challenge of finding a quantum mechanical approach to bonding. It should be apparent that the value of H_{12} and H_{21} should be related in some way to the bond length. If the two atoms are pulled completely apart, the nucleus of atom 1 would not attract the electron from atom 2 and vice versa.

The type of integral represented as S_{11} and S_{22} is called an *overlap integral*. These integrals give a view of how effectively the orbitals on the two atoms overlap. It should be clear that if the atomic wave functions are normalized,

$$S_{11} = S_{22} = \int \phi_1^* \phi_1 d\tau = \int \phi_2^* \phi_2 d\tau = 1. \tag{9.14}$$

Therefore, these integrals do not concern us much, but those of the type

$$S_{12} = \int \phi_1^* \phi_2 \, d\tau \tag{9.15}$$

$$S_{21} = \int \phi_2^* \phi_1 \, d\tau \tag{9.16}$$

do pose a problem. It should be inferred that these integrals represent the overlap of the wave function of the orbital from atom 1 with that of one from atom 2 and vice versa. Also, it is no surprise that the values of these integrals depend on the internuclear distance. If the two atoms were pushed closer and closer together until the two nuclei were at the same point, the overlap would be complete and the integrals S_{12} and S_{21} would be equal to 1. On the other hand, if the nuclei are pulled farther and farther apart, at a distance of infinity there would be no overlap of the atomic orbitals from atoms 1 and 2, and S_{12} and S_{21} would be equal to 0. Thus, the overlap integral must vary from 0 to 1 and its value must be a function of the distance between the two atoms. Furthermore, since the values of integrals of the type H_{12}, etc., are determined by bond length, it is apparent that it should be possible to find a relationship between H_{12} and S_{12}. As we shall see, this is precisely the case.

Making the substitutions for the integrals that appear in Eq. (9.7) leads to

$$E = \frac{a_1^2 H_{11} + 2a_1 a_2 H_{12} + a_2^2 H_{22}}{a_1^2 + 2a_1 a_2 S_{12} + a_2^2}. \tag{9.17}$$

Note that S_{11} and S_{22} have been omitted from the denominator in Eq. (9.17), their values being 1 if the atomic wave functions are normalized. It must be kept in mind that we are seeking the weighting parameters a_1 and a_2 introduced in Eq. (9.4) that make the energy a minimum, so we obtain these values by making use of

$$\left(\frac{\partial E}{\partial a_1}\right)_{a_2} = 0 \quad \text{and} \quad \left(\frac{\partial E}{\partial a_2}\right)_{a_1} = 0. \tag{9.18}$$

Differentiating Eq. (9.17) with respect to a_1 and simplifying gives Eq. (9.19). Repeating the differentiation with respect to a_2 gives Eqs. (9.19) and (9.20), which are written as

$$a_1 (H_{11} - E) + a_2 (H_{12} - S_{12}E) = 0 \tag{9.19}$$

$$a_1 (H_{21} - S_{21}E) + a_2 (H_{22} - E) = 0. \tag{9.20}$$

Equations (9.19) and (9.20), known as the *secular equations*, constitute a pair of linear equations of the form

$$\begin{aligned} ax + by &= 0 \\ cx + dy &= 0. \end{aligned} \tag{9.21}$$

Of course, it is obvious that the equations are satisfied if all of the coefficients are 0 (the *trivial solution*). A theorem of algebra, stated here without proof, requires that for a nontrivial solution, the determinant of the coefficients must be 0. Therefore,

$$\begin{vmatrix} a & b \\ c & d \end{vmatrix} = 0. \tag{9.22}$$

In the sular equations, the values of a_1 and a_2 are the unknown quantities that we seek to evaluate. Therefore, in determinant form the coefficients are

$$\begin{vmatrix} H_{11} - E & H_{12} - S_{12}E \\ H_{21} - S_{21}E & H_{22} - E \end{vmatrix} = 0. \quad (9.23)$$

This determinant is known as the *secular determinant.*

If the two atoms are identical (a homonuclear diatomic molecule), it is apparent that $H_{11} = H_{22}$ and $H_{12} = H_{21}$. Although it may not be quite as obvious, $S_{12} = S_{21}$ also and both of these integrals will be called simply S at this time. Therefore, expanding the determinant yields

$$(H_{11} - E)^2 - (H_{12} - SE)^2 = 0 \quad (9.24)$$

or

$$(H_{11} - E)^2 = (H_{12} - SE)^2. \quad (9.25)$$

Taking the square root of both sides of Eq. (9.25) gives

$$H_{11} - E = \pm(H_{12} - SE). \quad (9.26)$$

Solving this equation for E gives the values

$$E_b = \frac{H_{11} + H_{12}}{1 + S} \quad \text{and} \quad E_a = \frac{H_{11} - H_{12}}{1 - S}. \quad (9.27)$$

For these two energy states, E_b is referred to as the symmetric or bonding state and E_a is called the asymmetric or antibonding state. The combinations of the wave functions are shown graphically in Figure 9.4.

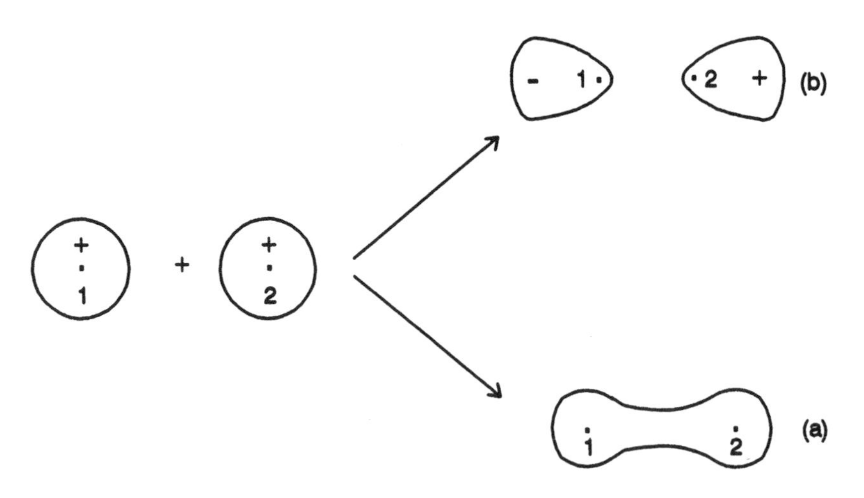

Figure 9.4 ▶ Electron density contours for (a) bonding and (b) antibonding σ orbitals.

Let us assume at this point that we have been considering the simplest molecule, H_2^+. The integral H_{11} represents the binding energy of the electron in the $1s$ state to nucleus 1 and H_{12} the additional attraction of the second nucleus to this negative charge. The integral H_{12} is the expression for the energy that results from the fact that the electron can also reside in the $1s$ state of the second hydrogen atom. As such, it is often called an *exchange integral*, and it also represents a negative energy or a binding energy. Consequently, the energy denoted as E_b represents the lower energy and the state is referred to as the bonding state. The state of higher energy, E_a, is called the antibonding state. The orbitals used in H_2 are identical to those used in H_2^+, although the energies are of course different. Figure 9.5 shows the molecular orbital diagram for the H_2 molecule.

Substituting the values for the energies shown in Eq. (9.27) into Eqs. (9.19) and (9.20) gives

$$a_1 = a_2 \quad \text{(symmetric state)} \tag{9.28}$$

$$a_1 = -a_2 \quad \text{(antisymmetric state).} \tag{9.29}$$

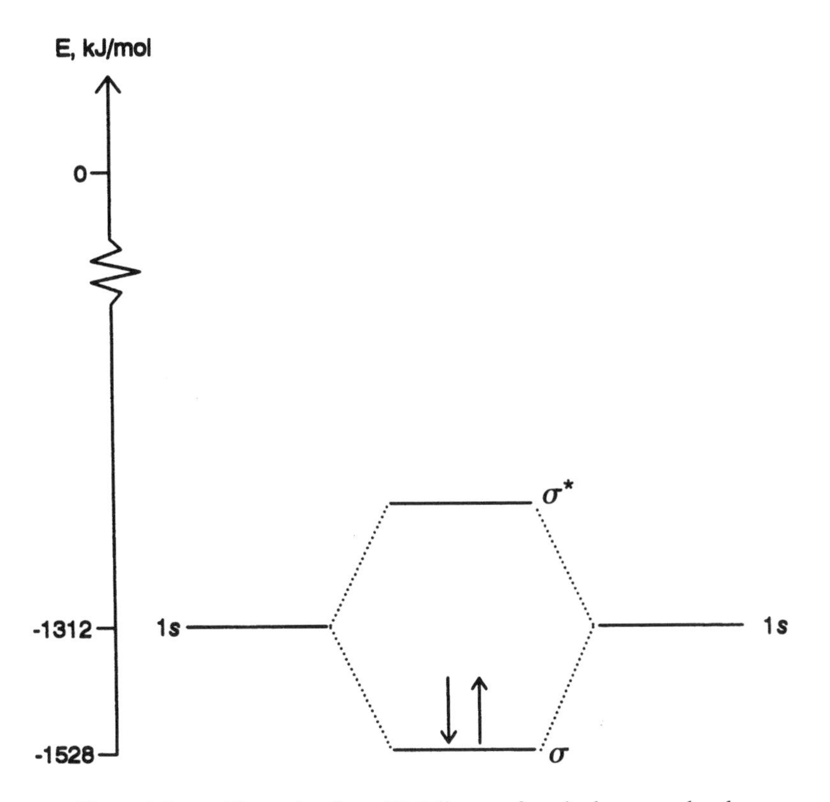

Figure 9.5 ▶ The molecular orbital diagram for a hydrogen molecule.

Consequently, the wave functions corresponding to the energy states E_b and E_a are

$$\psi_b = a_1\phi_1 + a_1\phi_2 = \frac{1}{\sqrt{(2+2S)}}(\phi_1 + \phi_2) \tag{9.30}$$

$$\psi_a = a_1\phi_1 - a_1\phi_2 = \frac{1}{\sqrt{(2-2S)}}(\phi_1 - \phi_2), \tag{9.31}$$

where the normalization constants represented as A are obtained as follows:

$$1 = \int A^2(\phi_1 + \phi_2)^2\, d\tau = A^2\left[\int \phi_1^2\, d\tau + \int \phi_2^2\, d\tau + 2\int \phi_1\phi_2\, d\tau\right]. \tag{9.32}$$

If the atomic wave functions ϕ_1 and ϕ_2 are normalized,

$$\int \phi_1\phi_2\, d\tau = 1$$

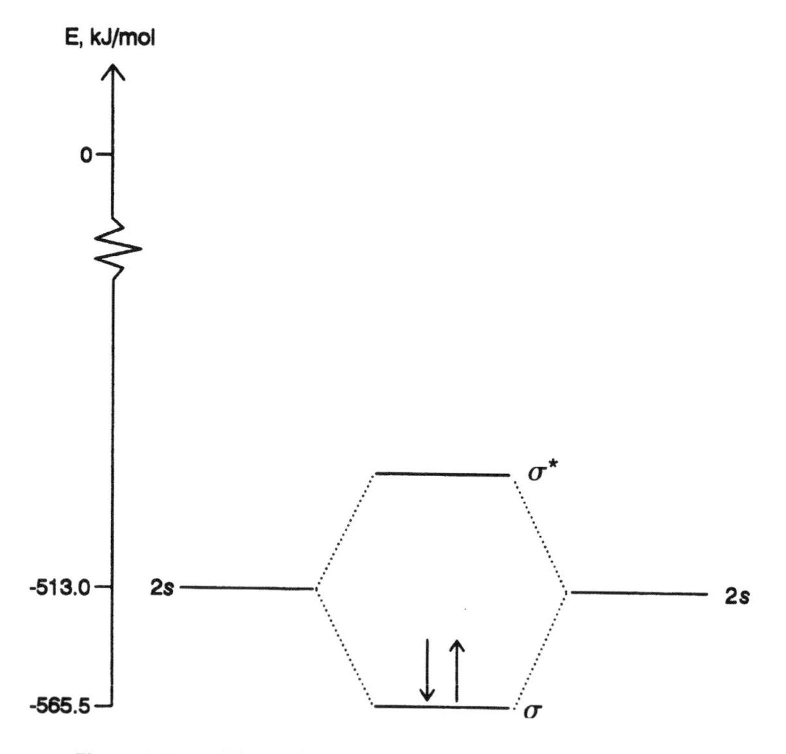

Figure 9.6 ▶ The molecular orbital diagram for lithium.

so that

$$1 = A^2 [1 + 1 + 2S]$$

and

$$A = \frac{1}{\sqrt{(2 + 2S)}}. \tag{9.33}$$

The combination of two 2s orbitals leads to the same type of molecular wave functions that we have already shown. Therefore, the molecular orbital diagram for Li_2, shown in Figure 9.6, has the same appearance as that for H_2 except that the energies of the atomic orbitals and the molecular orbitals are quite different.

9.4 Diatomic Molecules of the Second Period

In considering molecules composed of second-row atoms, the behavior of the p orbitals in forming molecular orbitals must be explored. For convenience, we take the direction lying along the bond to be the z axis. When the $2p_z$ orbitals overlap, a σ bond that is symmetric around the internuclear axis is formed. The wave functions produced by the combination of two $2p_z$ wave functions are

$$\psi\,(p_z) = \frac{1}{\sqrt{2 + 2S}}\,[\phi\,(z_1) + \phi\,(z_2)] \tag{9.34}$$

$$\psi\,(p_z)^* = \frac{1}{\sqrt{2 - 2S}}\,[\phi\,(z_1) - \phi\,(z_2)]\,. \tag{9.35}$$

After the p_z orbitals have formed a σ bond, the p_x and p_y orbitals form π bonds. The molecular orbital produced by the combination of two p_y orbitals produces a node in the xz plane, and the combination of the p_x orbitals produces a node in the yz plane. The combinations of atomic wave functions have the same form as those shown in Eqs. (9.34) and (9.35) except for the designation of the atomic orbitals. The electron density contours for these orbitals are symmetric on either side of the internuclear axis and represent π bonds. The electron density plots for the two π orbitals are shown in Figure 9.7.

The combination of the $2p$ wave functions also produces π and π^* orbitals, so we have from the combination of two sets of three p orbitals six molecular orbitals. The antibonding states lie higher in energy, but we still have a problem in ordering the energies of the σ and two π bonding

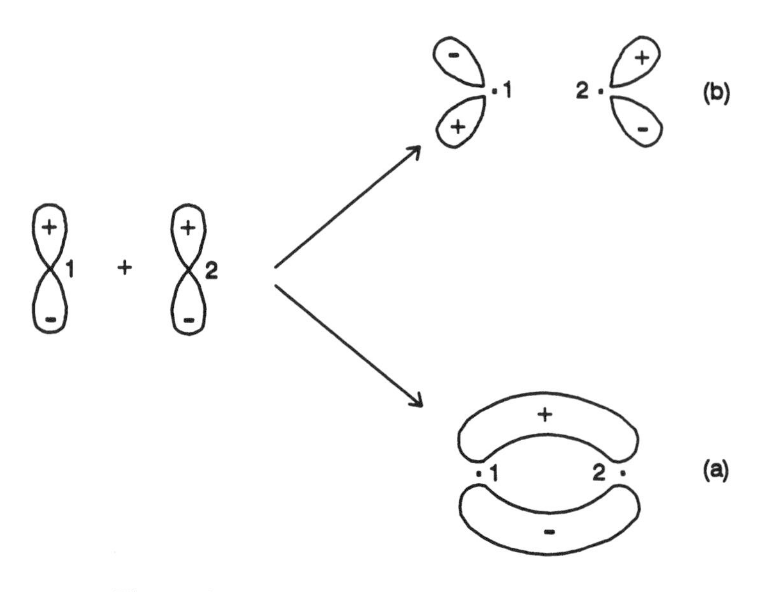

Figure 9.7 ▶ Electron density contours for (a) bonding and (b) antibonding π orbitals.

molecular orbitals. For some molecules, the σ orbital lies below the two degenerate π orbitals, while in other cases the two π orbitals lie lower in energy than the σ orbital. The reason for this is that for atoms early in the period (e.g., B and C), the $2s$ and $2p$ atomic orbitals are not greatly different in energy. This allows for partial mixing of the atomic states before the molecular orbitals form. In this case, the π orbitals are stabilized while the σ orbital is destabilized with the result that the two π orbitals lie lower in energy. For atoms later in the period (e.g., O and F), the higher nuclear charge causes the $2s$ orbital to lie well below the $2p$ orbitals so that there is no effective mixing of the atomic states. In these cases, the σ orbital lies lower in energy than the two π orbitals. For nitrogen, we might expect to have difficulty in deciding which arrangement of orbitals is correct, but the experimental evidence favors that in which the σ orbital lies lower than the π orbitals. Figure 9.8 shows the molecular orbital diagrams for the diatomic molecules of the second period that use $2p$ atomic orbitals.

Now that we have decided the order of orbitals with respect to increasing energy, we have only to fill them with the appropriate number of electrons according to Hund's rule. We have several experimental checks to see whether our predictions are correct. For example, the fact that the B_2 molecule has two unpaired electrons clearly indicates that the degenerate π orbitals lie below the σ orbital. The C_2 molecule is diamagnetic, but if

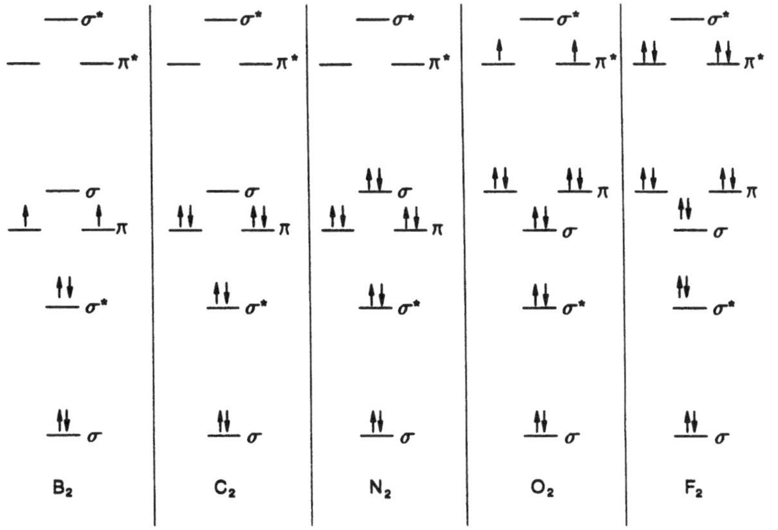

	B_2	C_2	N_2	O_2	F_2
Bond order	1	2	3	2	1
Bond energy (kJ/mol)	289	628	941	494	151

Figure 9.8 ▶ Molecular orbital diagrams for first-row diatomic molecules.

the σ orbital were lower in energy than the two π orbitals, it would be filled and the two π orbitals would hold one electron each. Therefore, in the C_2 molecule the π orbitals have lower energies than does the σ orbital. Moreover, the O_2 molecule has two unpaired electrons in the two π^* orbitals and it is paramagnetic. Figure 9.8 also gives values for the *bond order* (B), which is defined as

$$B = \frac{N_b - N_a}{2}, \tag{9.36}$$

where N_b is the number of electrons in bonding orbitals and N_a is the number of electrons in antibonding orbitals. Also shown in Figure 9.8 are the bond energies for the molecules, from which it is clear that the bond energy increases roughly with bond order.

Several important heteronuclear species are derived from the atoms of the second period. These include heteronuclear molecules like CO and NO and species like O_2^+ (dioxygenyl ion) and O_2^- (superoxide ion), for which molecular orbital diagrams are shown in Figure 9.9. Potassium reacts with

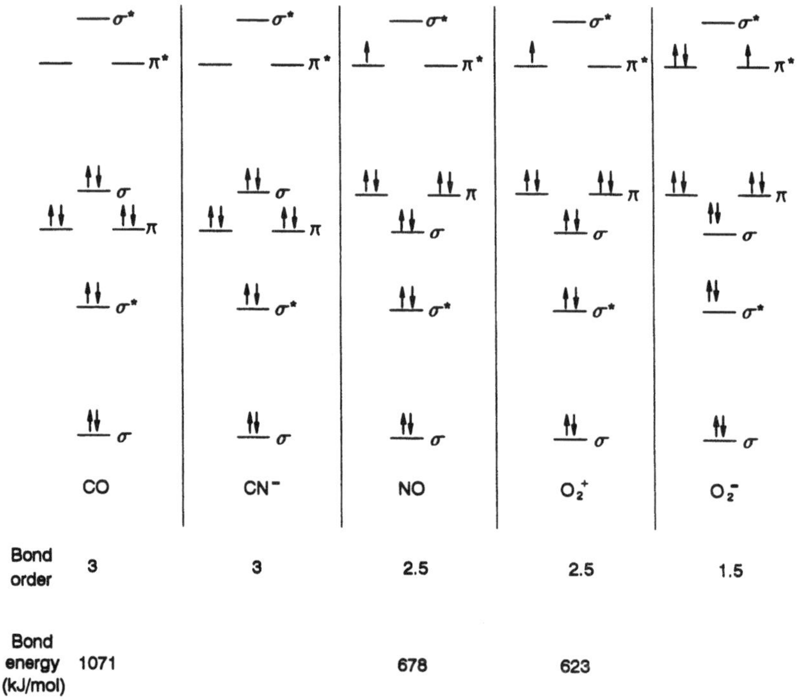

	CO	CN^-	NO	O_2^+	O_2^-
Bond order	3	3	2.5	2.5	1.5
Bond energy (kJ/mol)	1071		678	623	

Figure 9.9 ▶ Molecular orbital diagrams for diatomic ions and heteronuclear molecules of first-row elements.

oxygen to form O_2^-,

$$K + O_2 \longrightarrow KO_2, \tag{9.37}$$

while O_2^+ is generated by reaction of oxygen with PtF_6,

$$O_2 + PtF_6 \longrightarrow O_2^+ + PtF_6^-. \tag{9.38}$$

The electron lost from O_2 to produce the O_2^+ ion comes from an antibonding orbital, which results in an increase in the bond order from 2 to 2.5.

Another interesting species is the NO molecule, which has a single electron in a π^* orbital. In this case, the molecule can be ionized rather easily and the resulting NO^+ is then isoelectronic with CN^- and CO. Therefore, in some metal complexes NO behaves as a three-electron donor, one electron by way of ionization and an electron pair donated in the usual coordinate bond formation. The properties of many diatomic molecules are presented in Table 9.2.

TABLE 9.2 ▶ Properties for Diatomic Molecules and Ions

Molecule	N_b	N_a	B^a	R (Å)	DE[b] (eV)[c]
H_2^+	1	0	0.5	1.06	2.65
H_2	2	0	1	0.74	4.75
He_2^+	2	1	0.5	1.08	3.1
Li_2	2	0	1	2.62	1.03
B_2	4	2	1	1.59	3.0
C_2	6	2	2	1.31	5.9
N_2	8	2	3	1.09	9.76
O_2	8	4	2	1.21	5.08
F_2	8	6	1	1.42	1.6
N_2^+	7	2	2.5	1.12	8.67
O_2^+	8	3	2.5	1.12	6.46
BN	6	2	2	1.28	4.0
BO	7	2	2.5	1.20	8.0
CN	7	2	2.5	1.18	8.15
CO	8	2	3	1.13	11.1
NO	8	3	2.5	1.15	7.02
NO^+	8	2	3	1.06	—
SO	8	4	2	1.49	5.16
PN	8	2	3	1.49	5.98
SiO	8	2	3	1.51	8.02
LiH	2	0	1	1.60	2.5
NaH	2	0	1	1.89	2.0
PO	8	3	2.5	1.448	5.42

[a] B is the bond order.
[b] DE is the dissociative energy.
[c] One eV/molecule = 96.48 kJ/mol.

9.5 Overlap and Exchange Integrals

By now it should be apparent that the values of overlap integrals must be available if the molecular orbital energies are to be calculated. Over 50 years ago, Mulliken and co-workers evaluated the overlap integrals for a large number of cases using the Slater-type orbitals described in Chapter 5. Since orbital overlap is crucial to understanding covalent bonds, a convenient system of adjustable parameters is employed for determining overlap integrals. The parameters p and t are used in the Mulliken tables for atoms i and j, and they are defined as

$$p = 0.946R(\mu_i + \mu_j) \tag{9.39}$$

$$t = \frac{(\mu_i - \mu_j)}{(\mu_i + \mu_j)}. \tag{9.40}$$

In these equations, R is the internuclear distance in angstroms and μ is the quantity $(Z - S)/n$ from the Slater wave functions,

$$\psi = \left[R_{n,l}(r)\right] \exp\left[-(Z - S)\, r/a_0 n^*\right] Y(\theta, \phi). \tag{9.41}$$

When the spherical harmonics are combined with the radial function $R(r) = r^{n^*-1}$, the resulting wave function is written as

$$\psi = N r^{n^*-1} \exp(-\mu r/a_0), \tag{9.42}$$

where N is the normalization constant.

For atoms having $n \leq 3$, $n = n^*$ as was shown in the list of rules given in Chapter 5 for Slater wave functions. A brief list of μ values is shown in Table 9.3. If the calculations are for orbitals having different n values, the subscript i represents the orbital of smaller n. For cases where the orbitals have the same n, the i subscript represents the larger t value. The publication by Mulliken *et al.* includes extensive tables of values for the overlap integrals The tables include the orbital overlaps of $1s$, $1s$; $1s$, $2s$; $2s$, $2s$; etc., with the overlap integral being evaluated over a wide range of p and t values that cover a wide range of internuclear distances and types of atoms. These tables should be consulted if values for specific overlap integrals are needed. As we have seen earlier, the energy of the bonding molecular orbital can be written as

$$E_b = \frac{H_{11} + H_{12}}{1 + S}. \tag{9.43}$$

The value of H_{11} is obtained from the valence state ionization potential for the atom. If the molecular orbital holds two electrons, the bond energy (BE) is the difference between the energy of two electrons in the bonding state and their energy in the valence shells of the separated atoms. Therefore,

$$\text{BE} = 2H_{11} - 2\left(\frac{H_{11} + H_{12}}{1 + S}\right). \tag{9.44}$$

TABLE 9.3 ▶ Values of μ for Computation of Overlap Integrals

Atom	μ	Atom	μ
H	1.00	N	1.95
Li	0.65	O	2.275
Be	0.975	F	2.60
B	1.30	Na	0.733
C	1.625		

We can take the energy of the bond to be approximately $-2H_{12}/(1+S)$, and we need values for H_{12} and S. As we have seen, the integral H_{12} represents the exchange integral and is a negative quantity. The interaction that gives rise to this integral disappears at an infinite internuclear distance so $H_{12} = 0$ at $R = \infty$. However, we know that the overlap of the orbitals also becomes 0 when the atoms are separated by an infinite distance. Thus, we have two quantities that have values of 0 at infinity, but S becomes larger as the atoms get closer together while H_{12} becomes more negative under these conditions. What is needed is a way to express H_{12} as a function of the overlap integral. We also *suspect* that the exchange integral should be related to the energies with which electrons are bonded to atoms 1 and 2. These energies are the electron binding energies H_{11} and H_{22}.

As a first attempt, we will express H_{12} in terms of S and the average of the H_{11} and H_{22} integrals. Increasing S should increase H_{12}, and increasing the average of H_{11} and H_{22} should also increase H_{12}. By considering these factors, Mulliken assumed many years ago that the off-diagonal elements (H_{12}) should be proportional to the overlap integral. The function that results can be written as

$$H_{12} = -KS\left(\frac{H_{11} + H_{22}}{1+S}\right), \tag{9.45}$$

where K is a proportionality constant that has a numerical value of about 1.75 (a rather wide range of values has been used). This relationship is known as the *Wolfsberg–Helmholtz approximation* and is one of the most widely used approximations for the exchange integrals in molecular orbital calculations. Unfortunately, there is no clear agreement on the value to be used for K. Roald Hoffmann (1963) has given a detailed analysis of the factors involved in the choice of K value.

Because of the fact that atoms having considerably different ionization potentials must frequently be considered, the arithmetic mean used in the Wolfsberg–Helmholtz approximation may not be the best way to calculate an average. We saw in section 9.2 that the geometric mean is preferable in some instances. Therefore, we can write the exchange integral using the geometric mean of the ionization potentials as

$$H_{12} = -KS\,(H_{11} \cdot H_{22})^{1/2}. \tag{9.46}$$

This relationship is known as the *Ballhausen–Gray approximation*. We have previously mentioned that the bond energy can be written as $-2H_{12}/(1+S)$, so suddenly we see why the relationship given in Eq. (9.2), which gives the bond energy in terms of the geometric mean of the ionization potential, works so well! The intuitive approach used earlier is equivalent to the Ballhausen–Gray approximation.

A chemical bond between two atoms has an energy related to the bond length by a potential energy curve like that shown in Figure 7.5, p. 143. There is a minimum energy at the equilibrium distance. Neither of the expressions shown in Eqs. (9.45) and (9.46) goes through a minimum when the bond length varies. The value of H_{12} simply gets more negative as the bond length decreases because the value of S increases as the bond gets shorter. An approximation for H_{12} that does show a minimum as internuclear distance changes is that given by Cusachs,

$$H_{12} = \frac{1}{2} S(K - |S|)(H_{11} + H_{22}). \tag{9.47}$$

This function passes through a minimum with respect to bond length because S is a function of bond length and this function is a quadratic in S. This is in some ways the most satisfactory approximation to H_{12} of the three presented. However, the Wolfsberg–Helmholtz and Ballhausen–Gray approximations are still widely used.

9.6 Heteronuclear Diatomic Molecules

In the case of homonuclear diatomic molecules, the wave functions for the bonding and antibonding states were given by

$$\psi_{\mathrm{b}} = \frac{1}{\sqrt{2}} (\phi_1 + \phi_2) \tag{9.48}$$

and

$$\psi_{\mathrm{a}} = \frac{1}{\sqrt{2}} (\phi_1 - \phi_2) . \tag{9.49}$$

The contours of these molecular orbitals were found to be symmetrical about the center of the internuclear axis. In the case of heteronuclear diatomic molecules, this is not true since the bulk of the bonding orbital lies toward the atom having the higher electronegativity. Thus, the form of the molecular wave function is changed by weighting to take this difference into account. Accordingly, we write the wave functions as

$$\psi_{\mathrm{b}} = \phi_1 + \lambda\phi_2 \tag{9.50}$$

$$\psi_{\mathrm{a}} = \phi_1 - \lambda\phi_2, \tag{9.51}$$

where the parameter λ takes into account the difference in electronegativity of the two atoms.

The extent to which the atomic orbitals will combine to produce molecular orbitals depends upon the relative energies of the orbitals. The closer the energies of the orbitals on the two atoms, the more complete the "mixing" and the atomic orbitals lose their individuality more completely.

In considering the bonding in diatomic molecules composed of atoms of different types, there is an additional factor that must be kept in mind. This additional factor arises from the fact that more than one Lewis structure can be drawn for the molecule. In valence bond terms, when more than one acceptable structure can be drawn for a molecule or ion, the structures are called resonance structures and the actual structure is said to be a resonance hybrid of all the structures with appropriate weightings for each. As we shall soon describe, the *weighting coefficient*, λ, takes into account the contribution of the ionic structure to the wave function. Thus, the adjustable coefficients in a wave function serve the same purpose as being able to draw more than one valence bond structure for the molecule.

For a molecule AB, we can write structures

$$\underset{\text{I}}{A-B} \quad \underset{\text{II}}{A^+B^-} \quad \underset{\text{III}}{A^-B^+}.$$

Even for nonpolar molecules like H_2, the ionic structures contribute a substantial amount to the overall stability of the molecule. It can be shown that the added stability in the case of H_2 amounts to about 0.24 eV/molecule (23 kJ/mol). For the AB molecule described by the structures above, we can write the wave function as

$$\psi_{\text{molecule}} = a\psi_{\text{I}} + b\psi_{\text{II}} + c\psi_{\text{III}}, \tag{9.52}$$

where a, b, and c are constants and ψ_{I}, ψ_{II}, and ψ_{III} represent wave functions corresponding to resonance structures I, II, and III, respectively. For a molecule like H_2, the ionic structures contribute equally, but much less than the covalent structure so that the coefficients are related by $a \gg b = c$. For a heteronuclear molecule, one of the ionic structures is usually insignificant because it is unrealistic to expect the atom having the *higher* electronegativity to assume a positive charge and the atom having the *lower* electronegativity to assume a negative charge. For example, in HF the structures H–F and H^+F^- would contribute roughly equally, but the structure H^-F^+ is unrealistic owing to the much higher electronegativity of fluorine. Accordingly, we can neglect one of the structures for most heteronuclear diatomic molecules and write the wave function as

$$\psi_{\text{molecule}} = \psi_{\text{covalent}} + \lambda\psi_{\text{ionic}}, \tag{9.53}$$

where ψ_{ionic} corresponds to the ionic structure having the negative charge on the element of higher electronegativity. We are interested in obtaining the relative weightings of the covalent and ionic portions of the wave function for the molecule. This can be done by making use of dipole moments.

A purely covalent structure with equal sharing of the bonding electron pair would result in a dipole moment of 0 for the molecule. Likewise, a completely ionic structure in which an electron is transferred would result in a dipole moment, μ, equal to $e \cdot r$, where e is the charge on the electron and r is the internuclear distance. The ratio of the observed dipole moment to that calculated for the completely ionic structure gives the relative ionic character of the bond. Consider the data shown in Table 9.4 for the hydrogen halides. The percent ionic character is given as

$$\%\ \text{ionic character} = 100\mu_{\text{obs}}/\mu_{\text{ionic}}. \tag{9.54}$$

It should be recalled at this point that it is the square of the wave function that is related to the weighting given to the structure described by the wave function. Consequently, λ^2 gives the weighting given to the ionic structure and $1 + \lambda^2$ gives the weighting of the contribution from both the covalent and the ionic structures. Therefore, the ratio of the weighting of the ionic structure to the total is $\lambda^2/(1 + \lambda^2)$. Thus,

$$\%\ \text{ionic character} = \frac{100\lambda^2}{(1 + \lambda^2)}, \tag{9.55}$$

from which we obtain

$$\frac{\lambda^2}{(1 + \lambda^2)} = \frac{\mu_{\text{obs}}}{\mu_{\text{ionic}}}. \tag{9.56}$$

TABLE 9.4 ▶ Some Data for H*X* Molecules (*X* = a Halogen)

Molecule	r (Å)	μ_{obs} (D)	μ_{ionic} (D)	Fraction ionic character ($\mu_{\text{obs}}/\mu_{\text{ionic}}$)	$\chi_A - \chi_B$
HF	0.92	1.91	4.41	0.43	1.9
HCl	1.28	1.03	6.07	0.17	0.9
HBr	1.43	0.78	6.82	0.11	0.8
HI	1.62	0.38	7.74	0.05	0.4

Note. One debye (D) = 10^{-18} esu cm. χ_A and χ_B are electronegativities of A and B.

For HF, we know that the ratio of the observed dipole moment (1.91 D) to that for the hypothetical ionic structure (4.41 D) is 0.43. Therefore,

$$0.43 = \frac{\lambda^2}{1+\lambda^2}. \tag{9.57}$$

From this equation, we can solve for λ and we find that $\lambda = 0.87$. Therefore, for HF, the molecular wave function can be written as

$$\psi_{\text{molecule}} = \psi_{\text{covalent}} + 0.87\psi_{\text{ionic}}. \tag{9.58}$$

Based on this model, HF appears to be 43% ionic, so the actual structure can be considered as being a hybrid of 57% of the covalent structure and 43% of the ionic structure. This statement should not be taken too literally, but it does give an approach to the problem of polar molecules. A similar procedure shows that HCl is 17% ionic with $\lambda = 0.45$, HBr is 11% ionic with $\lambda = 0.36$, and HI is 5% ionic with $\lambda = 0.23$ according to this model. It was mentioned earlier that including the ionic structures H^+H^- and H^-H^+ for H_2 results in increased stability of the molecule. This effect is also manifested for molecules like HF. The difference between the actual bond dissociation energy and that predicted for the purely covalent bond provides a measure of this resonance stabilization energy. Although the purely covalent bond does not exist, there are two commonly used methods of estimating its energy that were proposed by Linus Pauling. The first of these methods makes use of the postulate of the arithmetic mean in which the energy of the hypothetical covalent bond between atoms A and B is taken as $(D_{AA} + D_{BB})/2$, where D_{AA} and D_{BB} are the bond dissociation energies of A_2 and B_2, respectively. Thus, the difference between the actual bond energy D_{AB} and $(D_{AA} + D_{BB})/2$ is the added stability of the bond, Δ, that results from the ionic resonance structure. Thus,

$$\Delta = D_{AB} - (D_{AA} + D_{BB})/2. \tag{9.59}$$

The quantity Δ, the resonance energy, is always positive since the actual bond energy is greater than predicted for the purely covalent bond alone.[1] Pauling realized that the extent to which an ionic structure stabilizes a diatomic molecule is related to a fundamental difference in the ability of the atoms to attract electrons. He therefore related Δ to the difference in this property, which we now recognize as the electronegativity, χ. Pauling's relationship is

$$\Delta = 23.06\,|\chi_A - \chi_B|^2. \tag{9.60}$$

[1]Strictly this is not entirely true. Pauling presents a discussion of cases like LiH, NaH, etc., for which Δ is negative (*The Nature of the Chemical Bond*, p. 82). However, such cases need not concern us now.

TABLE 9.5 ▶ Resonance Energies for Hydrogen Halides When the Arithmetic Mean Is Used to Estimate the Energy of the Covalent AB Bond

	HF	HCl	HBr	HI
D_{HX} (kJ/mol)	563	431	366	299
$(D_{HH} + D_{XX})/2$ (kJ/mol)	295	339	316	295
Δ (kJ/mol)	268	92	50	4
Electronegativity difference	1.9	0.9	0.7	0.4

Note. The bond energies used are H_2, 436; F_2, 153; Cl_2, 243; Br_2, 193; and I_2, 151 kJ/mol.

The values of Δ can now be considered as experimentally determined quantities and the electronegativities of A and B must be determined, but only the *difference* is known from Eq. (9.60). Obviously, the differences 100 − 99 and 3 − 2 are exactly the same. Pauling *assigned* the value for fluorine as 4.0 so that all other atoms had electronegativities between 0 and 4.0. The constant 23.06 merely converts eV/atom to kcal/mol since 1 eV/molecule = 23.06 kcal/mol. Table 9.5 summarizes similar calculations for all of the hydrogen halides.

Earlier in this chapter, we made use of the postulate of the geometric mean in determining the average ionization potential for two atoms. Pauling found also that the geometric mean gave better correlations of electronegativity and bond energy when the atoms have considerably different electronegativities. In making use of the geometric mean, the bond energy for the hypothetical purely covalent structure, D_{AB}, was taken as $(D_{AA} \cdot D_{BB})^{1/2}$. Using this approximation, the ionic stabilization, Δ, is given by

$$\Delta' = D_{AB} - (D_{AA} \cdot D_{BB})^{1/2}. \tag{9.61}$$

The values of Δ' are not linearly related to differences in electronegativity, but $(\Delta')^{1/2}$ is approximately a linear function of $|\chi_A - \chi_B|$. Table 9.6 shows the resonance stabilization for the hydrogen halides obtained using the geometric mean approximation.

There are several other equations that are sometimes used to estimate the percent ionic character of bonds in heteronuclear molecules. These are semiempirical equations that relate the percent ionic character to the difference in electronegativity for the atoms. Two of these equations are

$$\% \text{ ionic character} = 16\,|\chi_A - \chi_B| + 3.5\,|\chi_A - \chi_B|^2 \tag{9.62}$$

$$\% \text{ ionic character} = 18\,|\chi_A - \chi_B|^{1.4}. \tag{9.63}$$

TABLE 9.6 ▶ Resonance Energies for Hydrogen Halides Obtained Using the Geometric Mean for Calculating D_{AB}

	HF	HCl	HBr	HF
D_{HX} (kJ/mol)	563	432	366	299
$(D_{HH} \cdot D_{XX})^{1/2}$ (kJ/mol)	259	326	290	257
Δ'	304	106	76	42
$\sqrt{\Delta'}$	17.4	10.3	8.7	6.5
Electronegativity difference	1.9	0.9	0.7	0.4

Note. Bond energies used are the same as those listed in Table 9.5.

Although these equations appear quite different, the predicted percent ionic character is approximately the same for $|\chi_A - \chi_B|$ in the range of 1–2. Using these equations in combination with Eq. (9.55) enables an estimate of the weighting constant, λ, to be made if the electronegativities of the atoms are known.

Not only is there a contribution of the ionic structure to the bond energy, but it also leads to a shortening of the bond. For diatomic molecules, the bond length is usually given as the sum of the covalent radii of the atoms. However, if the atoms have different electronegativities, the sum of the covalent radii does not accurately give the observed internuclear distance because of the contribution of the ionic structure. A more accurate value for the bond length of a molecule *AB* that has some ionic character is given by the Shoemaker–Stevenson equation,

$$r_{AB} = r_A + r_B - 9.0\,|\chi_A - \chi_B|\,, \tag{9.64}$$

where r_A and r_B are the covalent radii of atoms *A* and *B*, respectively, and χ_A and χ_B are their electronegativities. In the molecule ClF, the observed bond length is 163 pm. The covalent radii for Cl and F are 99 and 72 pm, respectively, which leads to an expected bond length of 171 pm. Since the electronegativities of Cl and F are 3.0 and 4.0, Eq. (9.64) predicts a bond length of 162 pm for ClF, in excellent agreement with the experimental value. The experimental bond energy for ClF is 253 kJ/mol, which is considerably greater than the covalent value of 198 kJ/mol predicted from the arithmetic mean or 193 kJ/mol predicted using the geometric mean.

Combinations of atomic orbitals from atoms having different electronegativities (and hence different valence orbital energies) produce molecular orbitals that have energies closer to the atomic orbital of lower energy. In fact, the greater the electronegativity difference between the atoms, the closer the bond comes to being ionic. The bonding molecular orbital in that case represents an *atomic* orbital on the atom having the

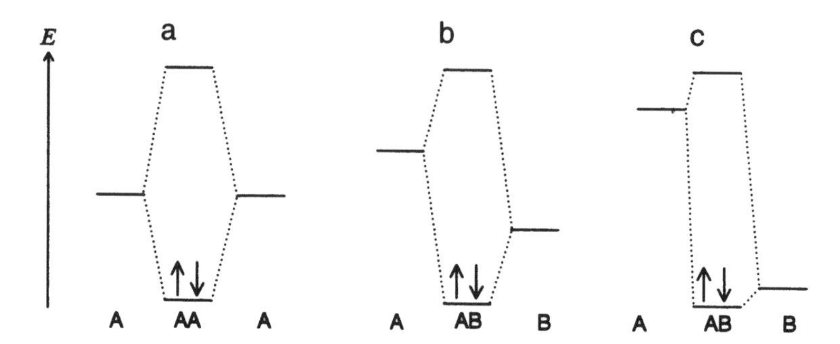

Figure 9.10 ▶ Molecular orbitals showing the effects of differences in electronegativity. (a) The two atoms have the same electronegativity. (b) Atom *B* has higher electronegativity. The molecular orbital has more of the character of an orbital of atom *B* (similar energy). (c) The difference in electronegativity is large enough that the electron pair essentially resides in an orbital on atom *B* (ionic bond).

higher electronegativity to which the electron is transferred. That is, the bond is essentially ionic. Figure 9.10 shows the change from equal sharing to electron transfer as the difference in electronegativity between the two atoms increases.

The concept of electronegativity gives us one of the most useful principles available when it comes to dealing with chemical bonding. It provides a measure of the ability of an atom in a molecule to attract electrons to itself. Therefore, we can predict bond polarities in most cases, although a few (e.g., CO) seem contradictory at first. Table 9.7 shows the electronegativities of the main group elements.

Because the basic idea regarding electron distribution in a bond is so useful, there have been numerous attempts to establish an electronegativity scale based on atomic properties rather than on bond energies as is the Pauling scale. One such scale is the Mulliken scale, which predicts the electronegativity of an atom from its ionization potential and electron affinity. Both of these properties are measures of the ability of an atom to attract an electron so it is natural to base electronegativity on them. Using this approach, the electronegativity EN is represented as

$$\mathrm{EN} = (\mathrm{IP} + \mathrm{EA})/2, \tag{9.65}$$

where IP is the ionization potential and EA is the electron affinity of the atom. The values for these properties are often expressed in electron volts while bond energies are expressed as kcal/mol or kJ/mol. To convert the Mulliken EN values to the equivalent Pauling values, the former are divided by 3.17. There have been numerous other scales devised to represent the

TABLE 9.7 ▶ Electronegativities of Atoms Based on the Pauling Approach

H 2.1									
Li 1.0	Be 1.6				B 2.0	C 2.5	N 3.0	O 3.5	F 4.0
Na 0.9	Mg 1.3				Al 1.5	Si 1.8	P 2.1	S 2.5	Cl 3.0
K 0.8	Ca 1.0	Sc 1.3	···	Zn 1.6	Ga 1.6	Ge 1.8	As 2.0	Se 2.4	Br 2.8
Rb 0.8	Sr 0.9	Y 1.1	···	Cd 1.7	In 1.7	Sb 1.8	Sb 1.9	Te 2.1	I 2.5
Cs 0.7	Ba 0.9	La 1.1	···	Hg 1.9	Tl 1.8	Pb 1.9	Bi 1.8	Po 1.9	At 2.2

electronegativities of atoms, but the Pauling scale is still the most widely used.

9.7 Symmetry of Molecular Orbitals

The formation of the H_2 molecule, which has a center of symmetry, gives rise to the combinations of atomic orbitals that can be written as $\phi_1 + \phi_2$ and $\phi_1 - \phi_2$. We will discuss the topic of symmetry in greater detail in Chapter 10. However, a center of symmetry is simply a point through which each atom can be moved to give the same orientation of the molecule. For a diatomic molecule like H_2 that point is the midpoint of the bond between the two atoms. It is equally valid to speak of a center of symmetry for wave functions. The first of the combinations of wave functions (as shown in Figure 9.4a) possesses a center of symmetry while the sond does not. Therefore, the $\phi_1 + \phi_2$ molecular wave function corresponds to the orbital written as σ_g while the combination $\phi_1 - \phi_2$ combination corresponds to σ_u^*. In these designations, "g" refers to the fact that the wave function retains the same sign when inflected through the center of symmetry and "u" indicates that the wave function changes sign when it is inflected through the center of symmetry. We say that the bonding orbital is symmetric and antibonding orbital is antisymmetric. However for π and π^* orbitals g and u refer to symmetry with respect to a plane that contains the internuclear axis (see Figure 9.6).

For diatomic molecules, the order of filling of molecular orbitals is σ, σ^*, (π, π), σ, (π^*, π^*), σ^* for the early part of the first long period) and $\sigma, \sigma^*, \sigma, (\pi, \pi), (\pi^*, \pi^*), \sigma^*$ for the latter part of the first long series. The designations (π, π) and (π^*, π^*) indicate pairs of degenerate molecular orbitals. The electron configuration for the hydrogen molecule can be shown as $(\sigma_g)^2$ and that for the C_2 molecule is designated as $(\sigma_g)^2(\sigma_u^*)^2(\pi_u)^2(\pi_u)^2$ (see Figure 9.8).

The molecular orbitals can be identified by applying labels that show the atomic orbitals that were combined to give them. For example, the orbital of lowest energy is $1s\sigma_g$ or $\sigma_g 1s$. Other orbitals would have designations like $2p_x\pi_u$, $2p_y\pi_g^*$, etc.

As a result of orbital mixing, a σ molecular orbital may not arise from the combination of pure s atomic orbitals. For example, we saw in Section 9.4 that there is substantial mixing of $2s$ and $2p$ orbitals in molecules like B_2 and C_2. Therefore, labels like $2s\sigma_g$ and $2s\sigma_u^*$ may not be strictly correct. Because of this, the molecular orbitals are frequently designated as

$$1\sigma_g, 1\sigma_u, 2\sigma_g, 2\sigma_u, 3\sigma_u, 1\pi_u, 1\pi_g, \ldots .$$

In these designations, the leading digit refers to the order in which an orbital having that designation is encountered as the orbitals are filled. For example, $1\sigma_g$ denotes the first σ orbital having g symmetry (a bonding orbital), $3\sigma_u$ means the third σ orbital having u symmetry (an antibonding orbital), etc. The asterisks on antibonding orbitals are not really needed since a σ orbital having u symmetry is an antibonding orbital and it is the antibonding π orbital that has g symmetry. Therefore, the g and u designations alone are sufficient to denote bonding or antibonding character. These ideas elaborate on those discussed in Section 9.4.

9.8 Orbital Symmetry and Reactivity

For approximately 50 years, it has been recognized that symmetry plays a significant role in the reactions between chemical species. In simple terms, many reactions occur because electron density is transferred (or shared) between species as the transition state forms. In order to interact favorably (to give an overlap integral greater than 0), it is necessary for the interacting orbitals to have the same symmetry (see Section 4.3).Otherwise, orthogonal orbitals give an overlap integral equal to 0. The orbitals involved in the interactions of reacting species are those of higher energy, the so-called *frontier orbitals*. These are the highest occupied molecular orbital (HOMO) and the lowest unoccupied molecular orbital (LUMO). As the species interact, electron density flows from the HOMO on one species to the LUMO on the other. In more precise terms, it can be stated that the orbitals must belong to the same symmetry type or point group for the orbitals to overlap with an overlap integral greater than 0.

As was described earlier in this chapter, orbitals of similar energy interact (overlap) best. Therefore, it is necessary that the energy difference between the HOMO on one reactant and the LUMO on the other be less than some threshold value for effective overlap to occur. As a reaction takes place, a bond in one reactant molecule is broken as another is being formed. When both orbitals are bonding orbitals, the bond being broken (electron density is being donated) is the one representing the HOMO in one reactant and the bond being formed is represented by the LUMO in the other (which is empty and receives electron density as the molecules interact). When the frontier orbitals are antibonding in character, the LUMO in one reactant molecule corresponds to the bond broken and the HOMO to the bond formed.

Consider the N_2 and O_2 molecules. The HOMOs of the N_2 molecule are π_u orbitals, which are antisymmetric, while the LUMOs of O_2 are half-filled π_g (or π_g^*). Therefore, interaction of the frontier orbitals of these two molecules is forbidden by symmetry. Electron density could flow from the HOMO (having g symmetry) of the O_2 molecule to the π_g orbital on N_2 except for the fact that oxygen has a higher electronegativity than nitrogen. Therefore, transfer of electron density from O_2 to N_2 is excluded on the basis of their chemical properties. The reaction

$$N_2(g) + O_2(g) \longrightarrow 2\,NO(g) \tag{9.66}$$

is accompanied by a high activation energy and is very difficult to carry out in accord with these observations.

In the reaction

$$H_2(g) + I_2(g) \longrightarrow 2\,HI(g) \tag{9.67}$$

we expect to find that electron density is transferred from hydrogen to iodine because of the difference in their electronegativities. However, the LUMO of I_2 is an antibonding orbital designated as σ_u (or σ_u^*) while the HOMO for H_2 is σ_g. Consequently, the overlap is zero for the HOMO of an H_2 molecule with the LUMO of an I_2 molecule and the expected interaction is symmetry forbidden as shown in Figure 9.11. Transfer of

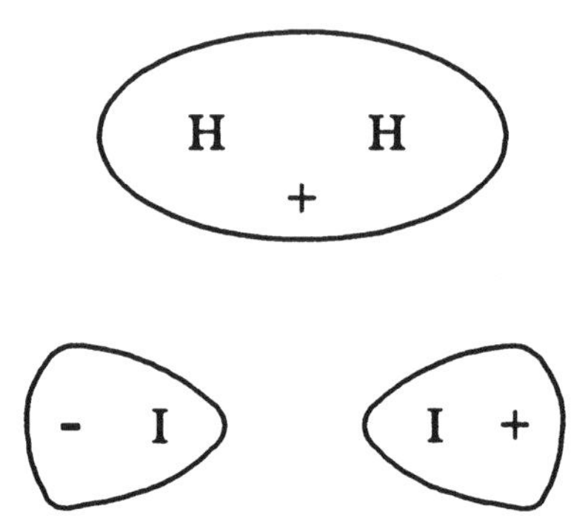

Figure 9.11 ▶ The interaction of hydrogen T_g and iodine T_u orbitals to form a bimolecular complex is symmetry forbidden.

electron density from filled molecular orbitals on I_2 to an empty molecular orbital on H_2 is not symmetry forbidden, but it is contrary to the difference in electronegativity. In view of these principles, it is not surprising that the reaction between H_2 and I_2 does not take place by a bimolecular process involving *molecules* although it was thought to do so for many years. A transition state like

$$\begin{array}{ccc} \mathrm{H} & \cdots & \mathrm{H} \\ \vdots & & \vdots \\ \mathrm{I} & \cdots & \mathrm{I} \end{array}$$

(where the dots indicate breaking and forming bonds) is forbidden on the basis of both chemical nature of the reactants and symmetry. This reaction actually occurs by the reaction of two iodine atoms with a hydrogen molecule, which is not symmetry forbidden.

9.9 Term Symbols

In Chapter 5, we discussed the spectroscopic states of atoms that result from the coupling of spin and orbital angular momenta. An analogous coupling of angular momenta occurs in molecules and for diatomic molecules; the coupling is similar to the Russell–Saunders scheme.

For a diatomic molecule, the internuclear axis is defined as the z axis (this will be discussed in greater detail in Chapter 10). In the case of atoms, we saw that the m value gave the projection of the l vector on the z axis. Also, the value of s gives the spin angular momentum in units of $\hbar$. To determine the spectroscopic ground state (indicated by a *term symbol*) for an atom, the sum of the spin angular momenta, S, and the sum of the orbital contributions, L, were obtained.

In a molecule, each electron has its own component of spin angular momentum and orbital angular momentum along the z axis. These angular momenta couple as they do in atoms, and the resultants determine the *molecular* term symbol. However, to represent the angular momenta of electrons in molecules, different symbols are used. The orbital angular momentum for an electron in a molecule is designated as λ. Also, the atomic orbitals combined to give a molecular orbital then have the same value for the z component of angular momentum. If the molecular orbital is a σ orbital, the orbital angular momentum quantum number is 0 so λ must equal 0 since that is the only projection on the z axis that a vector 0 units long can have. This means that for an electron in a σ orbital, the only m value possible is 0. For an electron in a π orbital, m can have values of $+1$ and -1, which are projections of a vector 1 unit in length. Therefore, if the molecular orbital is a π orbital, the λ value is 1. The total orbital angular momentum, Λ, is then determined. The resultant spin angular momentum is given by the sum of the electron spins as in the case of atoms. For atoms, the values $L = 0, 1$, and 2 give rise to spectroscopic states designated as S, P, and D, respectively. For molecules, the values $\Lambda = 0, 1$, and 2 give rise to spectroscopic states designated as Σ, Π, and Δ, respectively. Thus, coupling procedures for atoms and molecules are analogous except for the fact that Greek letters are used to denote molecular term symbols. The term symbol for a molecule is expressed as ${}^{2S+1}\Lambda$.

After the resultant spin and orbital angular momenta have been determined, these vectors can couple to give a total angular momentum. For atoms, the vector was designated as J, but for molecules it is usually designated as Ω, which has the possible values $0, 1, 2, \ldots$. As in the case of atoms, all filled shells have a total spin of 0 and the sum of the m values is also 0. Therefore, all of the lower lying filled shells can be ignored in determining the term symbol for a diatomic molecule.

In section 9.8, we saw that the molecular orbitals for homonuclear diatomic molecules are designated as g or u depending on whether they are symmetric or antisymmetric with respect to a center of symmetry. Of course, heteronuclear diatomic molecules do not possess a center of symmetry. The overall g or u character for more than one occupied orbital can be determined by considering g to represent a + sign and u to represent a − sign Then the g or u character of each orbital is multiplied by that of each other orbital. As a result, $g \times g = g$, $g \times u = u$, and $u \times u = g$. The σ states are also designated as Σ^+ or Σ^- based on whether the wave function that represents the molecular orbital is symmetric or antisymmetric with respect to reflection in any plane that contains the molecular axis.

For the H_2 molecule, the two electrons reside in the $1\sigma_g$ molecular orbital and the configuration is $(1\sigma_g)^2$. For a σ orbital, $\lambda = 0$ so the sum of the values for the two electrons is 0, which is the value for Λ. Therefore, the ground state is a Σ state and with the two electrons being paired, the sum of the spins is 0. The multiplicity, $2S + 1$, equals 1 so the ground state is $^1\Sigma$. Moreover, two electrons reside in the $1\sigma_g$ state and the product of g $\times$ g gives an overall symmetry of g. The molecular orbital wave function is symmetric with respect to a plane that contains the internuclear axis so the superscript + is appropriate. For the H_2 molecule, the correct term symbol is $^1\Sigma_g^+$ when all of the aspects are included.

When we work through the derivation of the term symbol for O_2, we begin with the electron configuration $(1\sigma_g)^2(1\sigma_u)^2(2\sigma_g)^2(1\pi_u)^4(1\pi_g)^2$, and the outer two electrons are unpaired in degenerate $1\pi_g$ orbitals. The sum of spins could be either 1 or 0, depending on whether the spins are aligned or opposed. The first of these values would give $2S + 1 = 3$ while the latter would give $2S + 1 = 0$. As we saw for atomic term symbols, the states of highest multiplicity would correspond to the lowest energy. This would occur when one electron resides in each orbital with parallel spins. However, this occurs when one electron is in each orbital so $m_{(1)} = +1$ and $m_{(2)} = -1$ so the sum is 0, which results in a Σ term. Therefore, the triplet state is a $^3\Sigma_g$ because both electrons reside in orbitals with g symmetry. Finally, the orbitals are symmetric with respect to a plane containing the internuclear axis so the superscript + is added to give $^3\Sigma_g^+$ as the term symbol. By drawing all of the microstates as was illustrated in section 5.4 we would find that other terms exist, but they do not represent the ground state.

Removal of an electron from the O_2 molecule to produce O_2^+ (the dioxygenyl ion) leaves one electron in a $1\pi_g$ orbital. In this case, $S = \frac{1}{2}$ and $\Lambda = 1$ and the molecular orbital has g symmetry, so the term symbol is $^2\Pi_g$. The cases that have been worked out here illustrate the procedures involved in finding the term of lowest energy for a particular configuration. However, as in the case of atoms, other terms are possible when all permissible microstates are considered.

References for Further Reading

- Coulson, C. A. (1961). *Valence*, 2nd ed. Oxford Univ. Press, New York. One of the classics on chemical bonding.
- Day, Jr., M. C., and Selbin, J. (1969). *Theoretical Inorganic Chemistry*, 2nd ed. Van Nostrand–Reinhold, New York. A good presentation of quantum mechanics applied to bonding.

- DeKock, R. L., and Gray, H. B. (1980). *Chemical Bonding and Structure*, Chap. 9. Benjamin–Cummings, Menlo Park, CA. An excellent, readable introduction to properties of bonds and bonding concepts.
- Eyring, H., Walter, J., and Kimball, G. E. (1944). *Quantum Chemistry*. Wiley, New York. High level and mathematical.
- Gray, H. B. (1965). *Electrons and Chemical Bonding*. Benjamin, New York. An elementary survey of bonding that contains a wealth of information.
- Hoffman, R. (1963). J. Chem. Phys. 39, 1397.
- Ketelaar, J. A. A. (1958). *Chemical Constitution*, 2nd ed. Elsevier, Amsterdam. Presents coverage of all types of interactions.
- Lowe, J. P. (1993). *Quantum Chemistry*, 2nd ed. Academic Press, San Diego. A standard higher level text on quantum chemistry. Chapter 14 includes a great deal of material on applications of molecular orbital theory to reactions.
- Mulliken, R. S., Rieke, A., Orloff, D., and Orloff, H. (1949). J. Phys. Chem. 17, 1248.
- Pauling, L. (1960). *The Nature of the Chemical Bond*, 3rd ed. Cornell Univ. Press, Ithaca, NY. A wealth of information on many topics.
- Pimentel, G. C., and Spratley, R. D. (1969). *Chemical Bonding Clarified through Quantum Mechanics*. Holden–Day, San Francisco, CA. An elementary book but one containing a great deal of insight.

Problems

1. For a molecule *XY*, $\psi_{\text{molecule}} = \psi_{\text{covalent}} + 0.50\psi_{\text{ionic}}$. Calculate the percent ionic character of the *X–Y* bond. If the bond length is 1.50 Å, what is the dipole moment?

2. For a homonuclear diatomic molecule, $H_{11} = H_{22}$ and $H_{12} = H_{21}$. This is not true for a heteronuclear molecule. Derive the expressions for the bonding and antibonding states for a heteronuclear diatomic molecule. You may still assume that $S_{12} = S_{21}$.

3. Suppose that the bond energies for A_2 and X_2 are 209 and 360 kJ/mol, respectively. If atoms *A* and *X* have electronegativities of 2.0 and 3.0, what will be the strength of the *A–X* bond? What will be the dipole moment if the internuclear distance is 1.25 Å?

4. Suppose that a diatomic molecule *XZ* contains a single σ bond. The binding energy of an electron in the valence shell of atom *X* is -10.0 eV. Spectroscopically it is observed that promotion of an electron to the antibonding state leads to an absorption band at $16{,}100\ \text{cm}^{-1}$. Using a value of 0.10 for the overlap integral, determine

the value of the exchange integral. Sketch the energy level diagram and determine the actual energies of the bonding and antibonding states. What is the bond energy?

5. For the molecule ICl, the wave function can be written as

$$\psi_{\text{ICl}} = \psi_{\text{covalent}} + 0.33\psi_{\text{ionic}}.$$

If the dipole moment for ICl is 0.65 D, what is the internuclear distance?

6. Write molecular orbital descriptions for NO, NO^+, and NO^-. Predict the relative bond energies of these species and account for any that are paramagnetic.

7. The covalent radii of F and Cl are 0.72 and 0.99 Å, respectively. Given that the electronegativities are 4.0 and 3.0, what would be the expected bond distance for ClF?

8. The H–S bond moment is 0.68 D and the bond length is 1.34 Å. What is the percent ionic character of the H–S bond?

9. Write the molecular orbital configurations for the following: (a) CO^+, (b) C_2^{2-}, (c) BO. Determine the bond order for these species. Write out the molecular orbital configurations for Na_2 and Si_2 (omit the $n = 1$ and $n = 2$ states and represent them as KK and LL). Give the bond order for each molecule. Describe the stabilities of these molecules when they are excited to the first excited state.

10. Calculate the percent ionic character for the following bonds. (a) HCl, (b) HC, (c) NH, and (d) LiH.

11. Give the term symbols for the boron atom in its ground state and arrange them in the order of increasing energy. What would be the spectroscopic state for boron in the first excited state?

12. A heteronuclear diatomic species has the molecular orbital electron configuration $1\sigma_g^2\ 1\sigma_u^2\ 2\sigma_g^2\ 2\sigma_u^2\ 1\pi_u^3$.

(a) What is the bond order for the species?

(b) Would the dissociation energy be higher or lower if an electron is removed from the $1\pi_u$ orbital? Explain.

(c) Would you expect this species to form a stable −1 ion?

13. What would be the symmetry designation (including g or u as appropriate) for the following atomic orbitals with respect to the z axis? (a) $2s$, (b) $2p_z$, (c) $2p_y$, and (d) $3d_{z^2}$.

14. Determine the ground-state term symbol for the following: (a) Li_2, (b) C_2, (c) O_2^-, and (d) B_2^+.

15. One term symbol for N_2^+ is $^2\Sigma_g^+$. From which orbital was the electron removed?

16. Explain the difference between the reactions

$$N_2 \longrightarrow N_2^+ + e^- \quad ^2\Sigma_g^+$$
$$N_2 \longrightarrow N_2^+ + e^- \quad ^2\Pi_u.$$

17. The bond energy in C_2^- is 2.2 eV greater than that of C_2 but the bond energy of O_2^- is 1.1 eV less than that of O_2. Explain this difference.

18. Explain why the bond length in F_2 is 142 pm while that of F_2^+ is 132 pm.

19. The ground state term for B_2 is $^3\Sigma_g^-$. Explain how this fact gives information about the order of filling the molecular orbitals for diatomic molecules of the sond row elements.

20. For the O_2 molecule, draw all of the microstates that could result for the $1\pi_g^1\ 1\pi_g^1$ configuration. Determine which term each belongs to. What is the ground state term?

Chapter 10

Symmetry

In Chapter 3, we saw that the symmetry (dimensions) of the box can affect the nature of the energy level diagram for a particle in a three-dimensional box model. Symmetry is, in fact, one of the most important and universal aspects of structures of all types. In chemistry and physics, symmetry and its application through group theory have a direct relationship to structures of molecules, combinations of atomic orbitals that form molecular orbitals, the types of vibrations possible for molecules, and several other facets of chemical sciences. For this reason, it is imperative that a brief introduction to symmetry be presented and its relationship to quantum mechanics be shown at an early level in the study of molecules.

10.1 What Symmetry Means

One of the most efficient ways to describe the spatial arrangement of atoms in a molecule is to describe the symmetry of the molecule. The symmetry of a molecule is denoted by a symbol that succinctly conveys the necessary information about how the atoms are arranged. The symbol used describes the *point group* to which the molecule belongs. Thus, the symbol O_h is used to describe a molecule like SF_6 having octahedral symmetry and C_{2v} is used to describe a bent molecule like H_2O. These symbols indicate the structures of the molecules having these symmetries. When they are encountered, these symbols denote to the reader a particular arrangement of atoms. For example, H_2O can have two orientations as shown in Figure 10.1. One orientation can be changed to the other by rotation of the molecule by 180°.

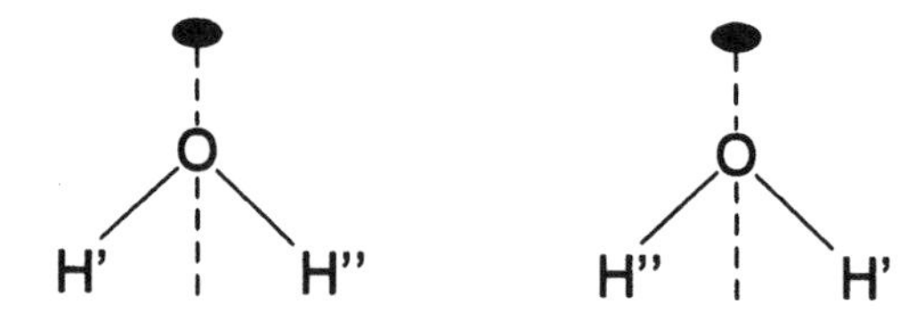

Figure 10.1 ▶ The orientation of the water molecule.

Figure 10.2 ▶ Two orientations of the HCl molecule.

On the other hand, the molecule HCl can have the two orientations shown in Figure 10.2. Rotation of HCl through 180° around the axis shown in the figure does not lead to an orientation indistinguishable from the original as it did in the case of H_2O. Thus, H_2O and HCl do not have the same symmetry. The details of the structure of a molecule can be adequately described if the symmetry properties are known and vice versa. The rules by which symmetry operations are applied and interpreted (group theory) give certain useful results for considering combinations of atomic wave functions as well. The rules of group theory govern the permissible ways in which the wave functions can be combined. Thus, some knowledge of molecular symmetry and group theory is essential to be able to understand molecular structure and molecular orbital theory. The discussion provided here will serve as only a brief introduction to the subject and the references at the end of the chapter should be consulted for a more exhaustive treatment.

As we have seen, a molecule may have two or more orientations in space that are indistinguishable. Certain parts of the molecule can be interchanged in position by performing some operation that changes the relative positions of atoms. Such operations are called *symmetry operations* and include rotations about axes and reflection through planes. The imaginary axes about which rotations are carried out are called *rotation axes*. The planes through which reflections of atoms occur are called *mirror planes*. Symmetry *elements* are the lines, planes, and points that relate the objects in the structure spatially. A symmetry *operation* is performed by making use of a symmetry element. Thus, symmetry elements and operations are not the same, but we will indicate a rotation axis by the symbol C and the operation of actually rotating the molecule around this axis by the same symbol. As a result, the axis is the C axis and the rotation about that axis is the C operation.

10.2 Symmetry Elements

Center of Symmetry or Inversion Center (i)

A molecule possesses a center of symmetry if inversion of each atom through this point results in an identical arrangement of atoms. Thus, in XeF_4 there is a center of symmetry:

```
      F
      |
  F—Xe—F
      |
      F
```

Of course each atom must be moved through the center the same distance that it was initially from the center. The Xe atom is at the center of symmetry and inversion of each fluorine atom through the Xe gives exactly the same arrangement as the original. However, for a tetrahedral CH_4 molecule, inversion through the *geometric center* of the molecule gives a different result:

```
    H              H
    |          H   |   H
    C            \ | /
  /  | \           C
 H   |   H         |
     H             H
```

Thus, the geometric center (where the C atom is located) is not a center of symmetry. Similarly, the linear CO_2 molecule has a center of symmetry, while the bent SO_2 molecule does not:

```
O=C=O          S
             //  \
            O     O
```

The (Proper) Rotation Axis (C_n)

If a molecule can be rotated around an imaginary axis to produce an equivalent orientation, the molecule possesses a *proper rotation axis*. We shall later describe an improper rotation axis (which is designated as S_n). Consider the boron trifluoride molecule, BF_3. As shown in Figure 10.3 we see

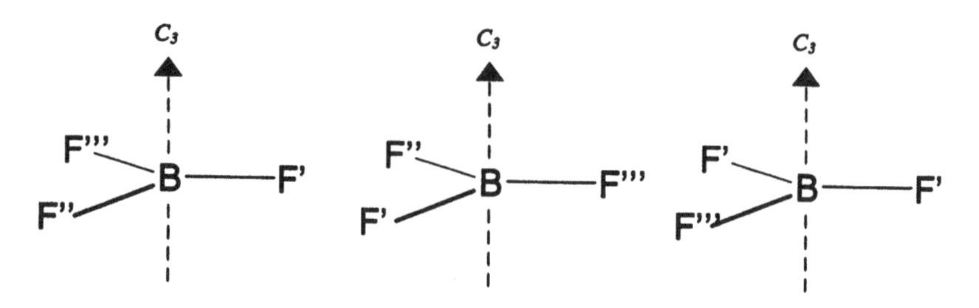

Figure 10.3 ▶ Rotation of a BF_3 molecule around a C_3 axis.

that rotation around an axis passing through the boron atom and perpendicular to the plane of the molecule produces an indistinguishable structure when the angle of rotation is 120°. In this case, the rotations producing indistinguishable orientations are 120°, or 360°/3, so that the rotation axis is a threefold or C_3 axis. Three such rotations return the fluorine atoms in the molecule to their original positions. For a C_n axis, the n value is determined by dividing 360° by the angle through which the molecule must be rotated to give an equivalent orientation. For the BF_3 molecule, there are three other axes about which the molecule can be rotated by 180° to arrive at the same orientation. These axes are shown in Figure 10.4. Therefore, there are three C_2 axes (one lying along each B–F bond) in addition to the C_3 axis that are perpendicular to the C_3 axis.

Although there are three C_2 axes, the C_3 axis is designated as the *principal* axis. The principal axis is designated as the axis of highest-fold rotation. This provides the customary way of assigning the z axis,

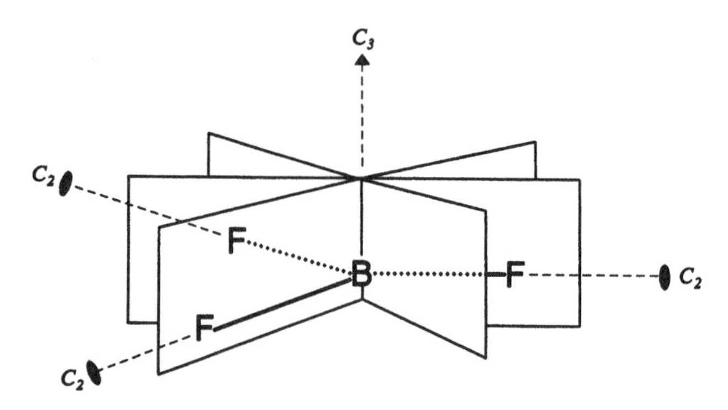

Figure 10.4 ▶ The trigonal planar BF_3 molecule showing the rotation axes and planes of symmetry.

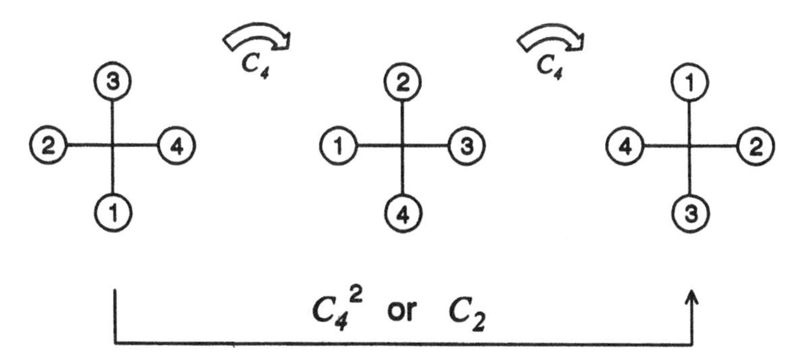

Figure 10.5 ▶ Rotation of a square planar molecule around a C_4 axis that is perpendicular to the plane of the page.

the axis of highest symmetry, in setting up an internal coordinate system for a molecule.

Multiple rotations are indicated as C_3^2, which means two rotations of 120° around the C_3 axis. This *clockwise* rotation of 240° produces the same orientation as a counterclockwise rotation of 120°. Such a rotation in the opposite direction is sometimes indicated as C_3^{-1}. It is readily apparent that the same orientation of BF_3 results from C_3^2 and C_3^{-1}.

Certain other "rules" can be seen by considering a square structure having four identical groups at its corners as shown in Figure 10.5. There is a C_4 axis perpendicular to the plane of the page at the center of the structure. If we perform four C_4 rotations, we arrive at a structure identical to the initial structure. Thus, C_n^n produces the *identity*, a structure indistinguishable from the original. The operation for the identity is designated as E so that $C_n^n = E$. It is also readily apparent that two rotations of 90° produce the same result as a single rotation of 180°, so that $C_4^2 = C_2$.

Mirror Plane (Plane of Symmetry) (σ)

If a molecule has a plane that divides it into two halves that are mirror images, the plane is known as a mirror plane (plane of symmetry). Consider the H_2O molecule shown in Figure 10.6. There are two mirror planes in this case, the xz and the yz planes. Reflection of the hydrogen atoms through the xz plane interchanges the locations of H′ and H″. Reflection through the yz plane simply interchanges the halves of the hydrogen atoms that are bisected by the yz plane. Both planes are designated as vertical planes because they encompass the z axis, which is the vertical axis.

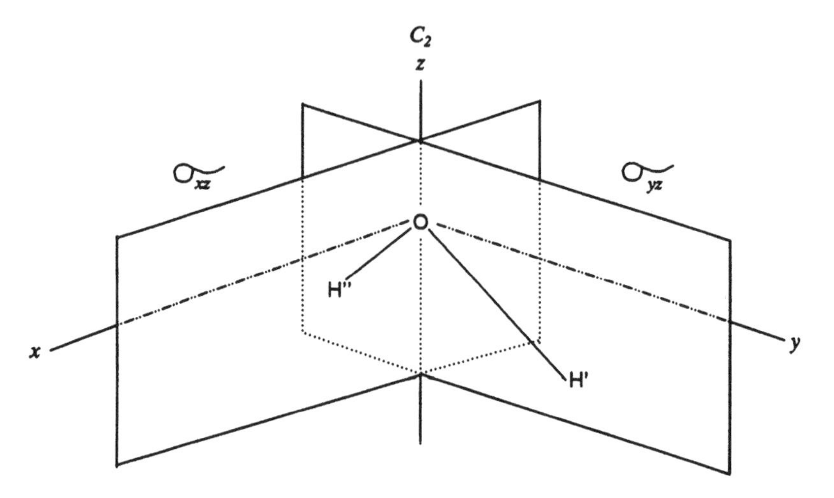

Figure 10.6 ▶ The water molecule showing two mirror planes. The intersection of these two planes generates a C_2 axis.

Improper Rotation Axis (S_n)

An improper rotation axis is one about which rotation followed by reflection through a plane perpendicular to the rotation axis produces an indistinguishable orientation. For example, the symbol S_6 means rotating the structure clockwise by 60° and reflecting each atom through a plane perpendicular to the S_6 axis of rotation. This can be illustrated with an example. Suppose we consider the six objects illustrated on the coordinate system shown in Figure 10.7. An open circle indicates a point below the xy

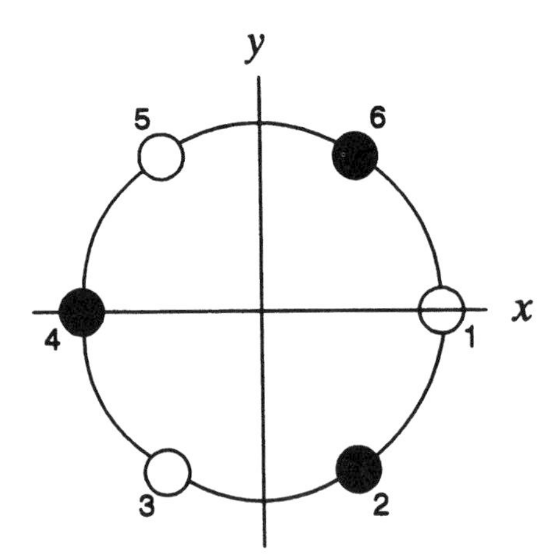

Figure 10.7 ▶ A structure possessing an S_6 (C_3) axis perpendicular to the plane of the page.

plane while a solid circle indicates a point lying above that plane (the plane of the page). The S_6 axis lies along the z axis, which is perpendicular to the x and y axes at the point of intersection. Rotation around the z axis by 60° followed by reflection through the xy plane moves the object at position 1 to position 2. Likewise, the object at position 2 moves to position 3, etc. Therefore, the S_6 operation has converted the original structure to another having the same orientation. It should be apparent that the zigzag or puckered structure shown in Figure 10.7 is the same as that of cyclohexane in the "chair" configuration. The S_6 axis is also a C_3 axis in this case because rotation by 120° around the z axis gives the same configuration:

From Figure 10.7 it should be clear that for this system

$$S_6^2 = C_6 \cdot \sigma_h \cdot C_6 \cdot \sigma_h = C_6^2 \cdot \sigma_h^2 = C_6 \cdot E = C_3$$

and that

$$S_6^4 = C_6 \cdot \sigma_h \cdot C_6 \cdot \sigma_h \cdot C_6 \cdot \sigma_h \cdot C_6 \cdot \sigma_h = C_6^4 \cdot E \cdot E = C_6^4 = C_3^2.$$

Consider a tetrahedral molecule as represented in Figure 10.8. In this case, it can be seen that rotation of the molecule by 90° around the z axis followed by reflection of each atom through the xy plane produces the structure in an unchanged orientation. The z axis is thus an S_4 axis for this molecule. It is easily seen that both the x and the y axes are S_4 axes also, so a tetrahedral molecule has as part of its symmetry elements three S_4 axes.

The Identity (E)

The identity operation can be carried out on all molecules of all symmetries since it leaves the orientation of the molecule unchanged. It is necessary to have the identity operation since an operation like C_n^n returns the molecule to its original orientation. Thus,

$$C_n^n = E.$$

Any symmetry operation (B) of the point group has an inverse operation (B^{-1}) such that

$$B \cdot B^{-1} = B^{-1} \cdot B = E.$$

We shall see the importance of the identity operation in Section 10.4 when elementary group theory is considered.

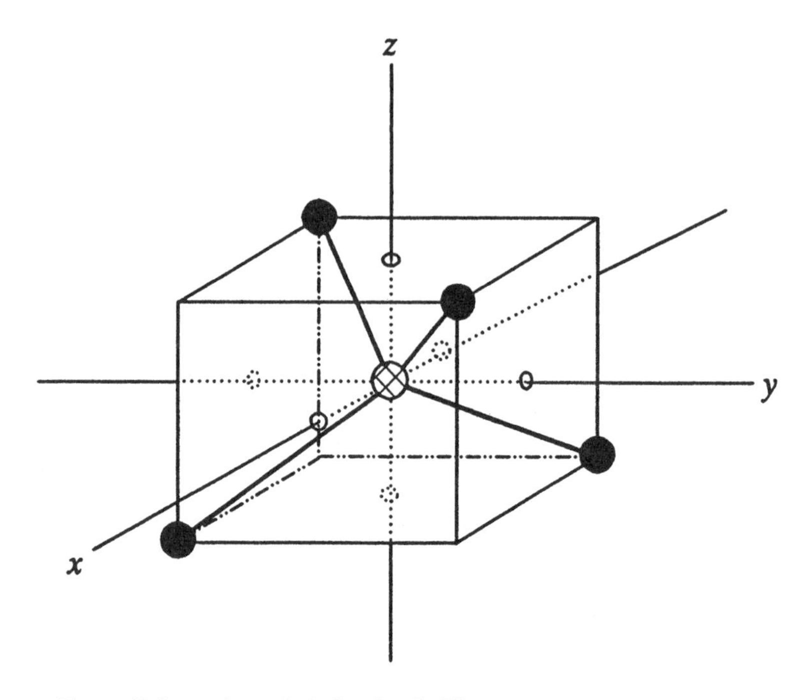

Figure 10.8 ▶ A tetrahedral molecule. The x, y, and z axes are S_4 axes.

The assignment of a molecule to a specific symmetry type (also called the *point group*) requires that the various symmetry elements present be recognized. This may not always be obvious, but with practice most molecules can readily be assigned to a symmetry type.

10.3 What Point Group Is It?

Determining the symmetry elements present in a molecule and then deducing the point group to which it belongs begins with drawing the *correct* structure. For example, H_2O is sometimes shown on printed pages as

$$\mathrm{H{-}\overline{\underline{O}}{-}H},$$

while the correct structure is

$$\begin{array}{ccc} & \mathrm{O} & \\ \diagup & & \diagdown \\ \mathrm{H} & & \mathrm{H} \end{array}$$

The incorrect linear structure appears to have a C_∞ axis because rotation of the molecule by any angle around a line along the bonds produces the same orientation. Also, a linear structure would possess a center of symmetry located at the center of the oxygen atom. Moreover, any line passing through the center of the oxygen atom and perpendicular to the C_∞ axis would be a C_2 axis. A plane perpendicular to the C_∞ axis cutting the oxygen atom in half would leave one hydrogen atom on either side and would be a mirror plane. Finally, a linear structure would have an infinite number of planes that bisect the molecule into equal fragments by cutting each atom in half, and the planes would intersect along the C_∞ axis.

A molecule possessing all these elements of symmetry is designated as having $D_{\infty h}$ symmetry or belonging to the $D_{\infty h}$ point group. It is easily seen from Figure 10.6 that the correct structure (bent or angular) has a C_2 axis and two mirror (vertical) planes that intersect along it, which are the only symmetry elements present. A molecule that has precisely these symmetry elements is called a C_{2v} molecule and belongs to the C_{2v} point group. Methane, CH_4, is sometimes shown on printed pages as

```
      H
      |
  H—C—H.
      |
      H
```

This structure would have a C_4 axis perpendicular to the plane of the page through the carbon atom with four vertical planes intersecting along it. Moreover, it would have a horizontal plane of symmetry, a center of symmetry, and four C_2 axes perpendicular to the C_4 axis. The point group for such a structure is D_{4h}. This is the correct structure for the planar XeF_4 molecule. Methane is actually tetrahedral,

```
      H
      |
      C    ,
    / | \
   H  H  H
```

and there are four C_3 axes and six mirror planes (not 12, each plane cleaves the molecule along two bonds). We have already shown that this molecule has three S_4 axes (see Figure 10.8). The point group for such a tetrahedral molecule is designated as T_d.

Table 10.1 summarizes the most common point groups of molecules and provides drawings showing the structures of the various types. Also, the molecular geometry is related to the hybrid orbital type of the central

TABLE 10.1 ► Symmetry of Molecules

Total number of electron pairs[a] and hybrid type	Number of unshared pairs of electrons[a]			
	0	1	2	3
2 sp	$D_{\infty h}$ Linear $BeCl_2$			
3 sp^2	D_{3h} Trigonal planar BCl_3	C_{2v} Bent $SnCl_2$		
4 sp^3	T_d Tetrahedral CH_4	C_{3v} Trigonal pyramid NH_3	C_{2v} Bent H_2O	
5 sp^3d	D_{3h} Trigonal bipyramid PCl_5	C_{2v} Irregular tetrahedron $TeCl_4$	C_{2v} "T" shaped ClF_3	$D_{\infty h}$ Linear ICl_2^-
6 sp^3d^2	O_h Octahedral SF_6	C_{4v} Square base pyramid IF_5	D_{4h} Square planar ICl_4^-	

[a] Around the central atom.

atom and the number of unshared pairs of electrons, if any. Table 10.2 provides a complete listing of the symmetry elements present in most of the commonly encountered point groups. Studying the structures in Table 10.1

TABLE 10.2 ▶ Common Point Groups and Their Symmetry Elements

Point group	Structure	Symmetry elements	Examples
C_1	—	None	CHFClBr
C_s	—	One plane	ONCl, $OSCl_2$
C_2	—	One C_2 axis	H_2O_2
C_{2v}	AB_2 bent or XAB_2 planar	One C_2 axis and two σ_v planes at 90°	H_2O, SO_2, NO_2, H_2CO
C_{3v}	AB_3 pyramidal	One C_3 axis and three σ_v planes	NH_3, PH_3, $CHCl_3$
C_{nv}	—	One C_n axis and n σ_v planes	BrF_5 (C_{4v})
$C_{\infty v}$	ABC linear	One C_∞ axis and ∞ σ_v planes	HCN, SCO, OCN^-, SCN^-
D_{2h}	Planar	Three C_2 axes, one σ_h and two σ_v planes, and center of symmetry	C_2H_4, N_2O_4
D_{3h}	AB_3 planar	One C_3 and three C_2 axes, and one σ_h and three σ_v planes	BF_3, CO_3^{2-}, NO_3^-, SO_3
D_{4h}	AB_4 planar	One C_4 and four C_2 axes, one σ_h and four σ_v planes, and center of symmetry	XeF_4, $PtCl_4^{2-}$
$D_{\infty h}$	AB_2 linear	One C_∞ axis, one σ_h and ∞ σ_v planes, and center of symmetry	CO_2, NO_2^+, CS_2
T_d	AB_4	Four C_3, three C_2, and three S_4 axes, and six σ_v planes	CH_4, P_4, MnO_4^-, SO_4^{2-}
O_h	AB_6 octahedral	Three C_4, four C_3, six C_2, four S_6, and three S_4 axes, nine σ_v planes, and center of symmetry	SF_6, $Cr(CO)_6$, PF_6^-
I_h	Icosahedral	Six C_5, 10 C_3, 15 C_2 axes, and 20 S_6 axes, and 15 planes	B_{12}, $B_{12}H_{12}^{2-}$

and the listing of symmetry elements in Table 10.2 should make it easy to assign the point group for most molecules, assuming that the structure is drawn correctly in the first place!

10.4 Group Theory

The mathematical apparatus for treating combinations of symmetry operations lies in the branch of mathematics known as group theory. A group is a set of elements and the corresponding operations that obey the following rules:

1. The combination of any two members of a group must yield another member of the group (*closure*).
2. The group contains the identity, E, multiplication by which commutes with all other members of the group ($EA = AE$) (*identity*).
3. The associative law of multiplication must hold so that $(AB)C = A(BC) = (AC)B$ (*associative*).
4. Every member of the group has a reciprocal such that $B \cdot B^{-1} = B^{-1} \cdot B = E$, where the reciprocal is also a member of the group (*inverse*).

Let us illustrate the use of these rules by considering the structure of the water molecule shown earlier in Figure 10.6. First, it should be apparent that reflection of the hydrogen atoms through the xz plane, indicated by σ_{xz}, transforms H′ into H″. More precisely, we could say that H′ and H″ are interchanged by reflection through the xz plane. Since the z axis coincides with a C_2 rotation axis, rotation by 180° about the z axis of the molecule will take H′ into H″ and H″ into H′, but with the "halves" of each interchanged with respect to the yz plane. The same result would follow from reflection through the xz plane followed by reflection through the yz plane. Therefore, in terms of operations

$$\sigma_{xz} \cdot \sigma_{yz} = C_2 = \sigma_{yz} \cdot \sigma_{xz},$$

where C_2 is rotation around the z axis by 360°/2. This establishes that σ_{xz} and σ_{yz} are both members of the group for this molecule. We see that in accord with rule 1 the combination of two members of the group has produced another member of the group, C_2.

If reflection through the xz plane is followed by repeating that operation, the molecule ends up with the arrangement shown in Figure 10.6. Symbolically,

$$\sigma_{xz} \cdot \sigma_{xz} = E.$$

Also, it is easy to see from Figure 10.6 that

$$\sigma_{yz} \cdot \sigma_{yz} = E$$

and

$$C_2 \cdot C_2 = E.$$

Further examination of Figure 10.6 shows that reflection of the molecule through the yz plane, σ_{yz}, will cause the “halves” of the H′ and H″ atoms lying on either side of the yz plane to be interchanged. If we perform that operation and then rotate the molecule by 360°/2 around the C_2 axis, we achieve exactly the same result as reflection through the xz plane produces. Thus,

$$\sigma_{yz} \cdot C_2 = \sigma_{xz} = C_2 \cdot \sigma_{yz}.$$

In a similar way, it is easy to see that reflection through the xz plane followed by a C_2 operation gives the same result as σ_{yz}. Finally, it can be seen from Figure 10.6 that reflections σ_{xz} and σ_{yz} in either order give the same orientation as results from the C_2 operation. Therefore

$$\sigma_{xz} \cdot \sigma_{yz} = C_2 = \sigma_{yz} \cdot \sigma_{xz}.$$

The associative law, rule 3, has also been demonstrated in these operations.

Additional relationships are provided by

$$E \cdot E = E$$
$$C_2 \cdot E = C_2 = E \cdot C_2$$
$$\sigma_{yz} \cdot E = \sigma_{yz} = E \cdot \sigma_{yz}, \text{ etc.}$$

All of these combinations of operations can be summarized in a *group multiplication table.*

$vThemultiplicationtable(Table$ 10.3) $fortheC_{2v}$, group is constructed so that the combination of operations follows the four rules presented at the beginning of this section. Obviously, a molecule having a structure other than C_{2v} (symmetry elements and operations) would require a different table.

To provide further illustrations of the use of symmetry elements and operations, the ammonia molecule, NH_3, will be considered. Figure 10.9

TABLE 10.3 ▶ Multiplication of Symmetry Operations for H_2O (C_{2v})

	E	C_2	σ_{xz}	σ_{yz}
E	E	C_2	σ_{xz}	σ_{yz}
C_2	C_2	E	σ_{yz}	σ_{xz}
σ_{xz}	σ_{xz}	σ_{yz}	E	C_2
σ_{yz}	σ_{yz}	σ_{xz}	C_2	E

Note. Start with an operation in the extreme left column and proceed to the column at whose top the desired operation is shown.

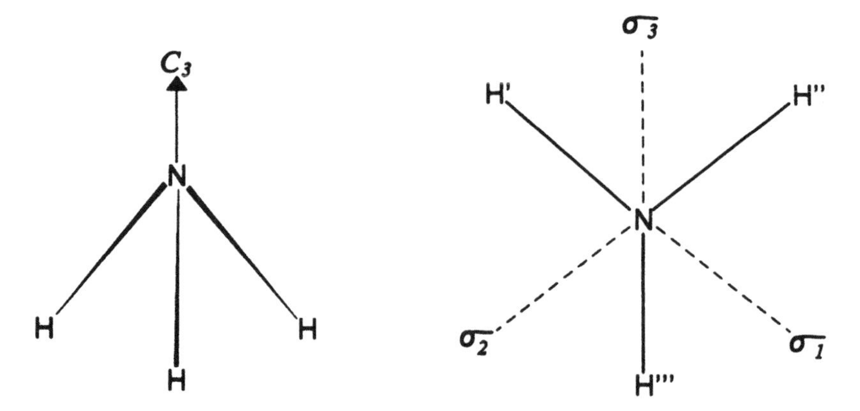

Figure 10.9 ▶ The pyramidal ammonia (C_{3v}) molecule. In the structure shown on the right, the C_3 axis is perpendicular to the page at the nitrogen atom.

shows that the NH_3 molecule has a C_3 axis passing through the nitrogen atom and three reflection planes containing that C_3 axis. The identity operation, E, and the C_3^2 operation complete the list of symmetry operations for the NH_3 molecule. Using the procedures illustrated above, we can write

$$C_3 \cdot C_3 = C_3^2$$
$$C_3^2 \cdot C_3 = C_3 \cdot C_3^2 = E$$
$$\sigma_1 \cdot \sigma_1 = E = \sigma_2 \cdot \sigma_2 = \sigma_3 \cdot \sigma_3.$$

Reflection through σ_2 does not change H$''$ but does interchange H$'$ and H$'''$. Reflection through σ_1 leaves H$'$ in the same position but interchanges H$''$ and H$'''$. We can summarize these operations as

$$\mathrm{H}' \overset{\sigma_2}{\leftrightarrow} \mathrm{H}'''$$
$$\mathrm{H}''' \overset{\sigma_1}{\leftrightarrow} \mathrm{H}''.$$

However, operation C_3^2 would move H$'$ to H$'''$, H$''$ to H$'$, and H$'''$ to H$''$, which is exactly the same orientation as that produced when σ_2 is followed by σ_1. It follows, therefore, that

$$\sigma_2 \cdot \sigma_1 = C_3^2.$$

This process could be continued so that all the combinations of symmetry operations would be worked out. Table 10.4 shows the resulting multiplication table for the C_{3v} point group, which is the point group to which a pyramidal molecule like NH_3 belongs.

TABLE 10.4 ▶ The Multiplication Table for the C_{3v} Point Group

	E	C_3	C_3^2	σ_1	σ_2	σ_3
E	E	C_3	C_3^2	σ_1	σ_2	σ_3
C_3	C_3	C_3	E	σ_3	σ_1	σ_2
C_3^2	C_3^2	E	C_3	σ_2	σ_3	σ_1
σ_1	σ_1	σ_2	σ_3	E	C_3	C_3^2
σ_2	σ_2	σ_3	σ_1	C_3^2	E	C_3
σ_3	σ_3	σ_1	σ_2	C_3	C_3^2	E

TABLE 10.5 ▶ Character Table for the C_{2v} Point Group

	E	C_2	σ_{xz}	σ_{yz}
A_1	1	1	1	1
A_2	1	1	−1	−1
B_1	1	−1	1	−1
B_2	1	−1	−1	1

Multiplication tables can be constructed for the combination of symmetry operations for a large number of other point groups. However, it is not the multiplication table as such that is of interest. Let us return to the multiplication table for the C_{2v} point group given in Table 10.3.

The symbols at the left in Table 10.5 give the symmetry properties of the irreducible *representation* of the C_{2v} group. We will now discuss in an elementary way what the symbols mean.

Suppose we have a vector of unit length lying coincident with the x axis as shown in Figure 10.10. The identity operation does not change the orientation of the vector. Reflection in the xz plane leaves the vector unchanged but reflection through the yz plane changes it to a unit vector in the $-x$ direction. Likewise, the C_2 operation around the z axis changes the vector so it points in the negative direction. Therefore, the vector is said to *transform* as $+1$ for the operations E and σ_{xz} but to transform as -1 for the operations C_2 and σ_{yz}. Table 10.5 shows the row containing the numbers $+1, -1, +1$, and -1 under the operations E, C_2, σ_{xz}, and σ_{yz}, respectively, labeled as B_1. It is easy to show how the numbers in the other rows can be obtained in a similar manner. The four representations, A_1, A_2, B_1, and B_2, are known as the *irreducible representations* of the C_{2v} group. It can be shown that these four irreducible representations cannot be separated or decomposed into other representations.

For a given molecule belonging to a particular point group, it is possible to consider the A_1, A_2, B_1, and B_2 symmetry species as indicating the behavior of the molecule under symmetry operations. As will be shown

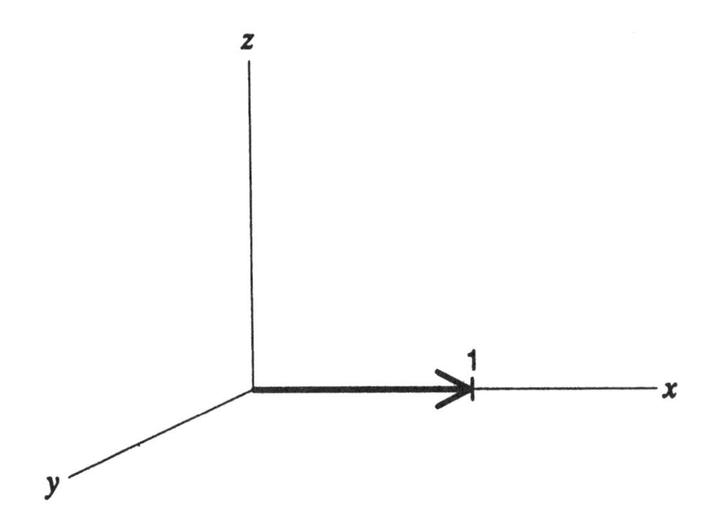

Figure 10.10 ▶ A unit vector lying on the x axis.

later, similar species also determine the ways in which the atomic orbitals can combine to produce molecular orbitals since the combinations of atomic orbitals must satisfy the character table of the group. We need to give some meaning that is based on the molecular structure for the species A_1, B_2, etc.

The following conventions are used to label species in the character tables corresponding to the various point groups:

1. The symbol A is used to designate a nondegenerate species symmetric about the principle axis.

2. The symbol B is used to designate a nondegenerate species antisymmetric about the principle axis.

3. The symbols E and T represent doubly and triply degenerate species, respectively.

4. If a molecule possesses a center of symmetry, the letter g indicates symmetry with respect to that center (*gerade*), and the letter u indicates antisymmetry with respect to that center of symmetry (*ungerade*).

5. For a molecule that has a rotation axis other than the principal one, symmetry or antisymmetry with respect to that axis is indicated by a subscript 1 or 2, respectively. When no rotation axis other

than the principal one is present, these subscripts are sometimes used to indicate symmetry or antisymmetry with respect to a vertical plane, σ_v.

6. The marks $'$ and $''$ are sometimes used to indicate symmetry or antisymmetry with respect to a horizontal plane, σ_h.

It should now be apparent how the species A_1, A_2, B_1, and B_2 arise. Character tables have been worked out and are tabulated for all of the common point groups. Presenting all the tables here would go beyond the scope of this introductory book.

We have barely scratched the surface of the important topic of symmetry. An introduction such as that presented here serves to introduce the concepts and nomenclature as well as making one able to recognize the more important point groups. Thus, the symbol T_d or D_{4h} takes on precise meaning in the language of group theory. The applications of group theory include, among others, coordinate transformations, analysis of molecular vibrations, and the construction of molecular orbitals. Only the last of these uses will be illustrated here.

10.5 Construction of Molecular Orbitals

The application of symmetry concepts and group theory greatly simplifies the construction of molecular wave functions from atomic wave functions. For example, it can be shown that the combination of two hydrogen $1s$ wave functions $(\phi_{1s(1)} + \phi_{1s(2)})$ transforms as A_1 (sometimes written as a_1 when *orbitals* are considered), and the combination $(\phi_{1s(1)} - \phi_{1s(2)})$ transforms as B_1 (sometimes written as b_1). According to the description of species given above, we see that the A_1 combination is a singly degenerate state symmetric about the internuclear axis. Also, the B_1 combination represents a singly degenerate state antisymmetric about the internuclear axis. The states represented by the combinations $(\phi_{1s(1)} + \phi_{1s(2)})$ and $(\phi_{1s(1)} - \phi_{1s(2)})$ describe the bonding (A_1) and antibonding (B_1) molecular states, respectively, as shown in Figure 10.11, where a_1 and b_1 are used to denote the orbitals.

It can be shown that for any group, the irreducible representations must be orthogonal. Therefore, only interactions (combinations) of orbitals having the same irreducible representations lead to nonzero elements in the secular determinant. It remains, then, to determine how the various orbitals transform under different symmetry groups. For the H_2O molecule, the coordinate system is shown in Figure 10.6. Performing any of the four

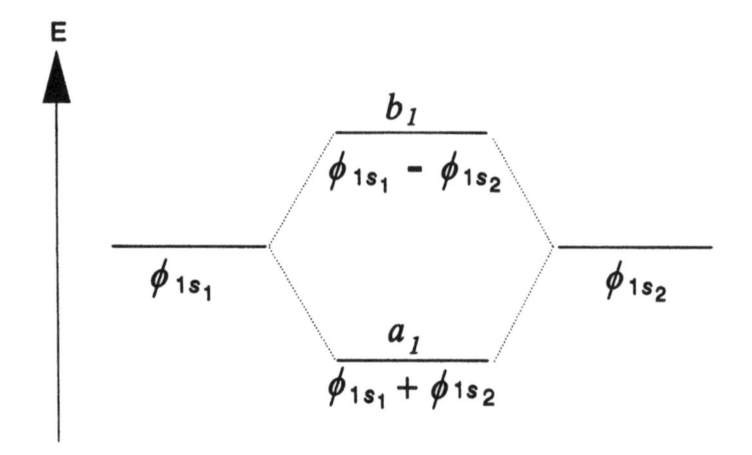

Figure 10.11 ▶ Two combinations of $1s$ wave functions giving different symmetry.

operations possible for the C_{2v} group (E, C_2, σ_{xz}, and σ_{yz}) leaves the $2s$ orbital unchanged. Therefore, that orbital transforms as A_1 (values of $+1, +1, +1$, and $+1$). Likewise, the p_x orbital does not change sign under E or σ_{yz} operations, but it does change signs under C_2 and σ_{yz} operations. This orbital thus transforms as B_1 ($+1, -1, +1$, and -1). In a like manner, we find that p_z transforms as A_2 (does not change signs under C_2, E, σ_{xz}, or σ_{yz} operations). Although it may not be readily apparent, the p_y orbital transforms as B_2. Using the four symmetry operations for the C_{2v} point group, we find that the valence shell orbitals of oxygen behave as follows:

Orbital	*Summary*
$2s$	A_1
$2p_z$	A_2
$2p_x$	B_1
$2p_y$	B_2

The possible wave functions constructed for the molecular orbitals in molecules are those constructed from the irreducible representations of the groups giving the symmetry of the molecule. These are readily found in the character table for the appropriate point group for the molecule. For the water molecule, which has the point group C_{2v}, the character table (see Table 10.5) shows that only A_1, A_2, B_1, and B_2 representations occur for a molecule having C_{2v} symmetry. We can use this information to construct a qualitative molecular orbital scheme for the H_2O molecule as shown in Figure 10.12.

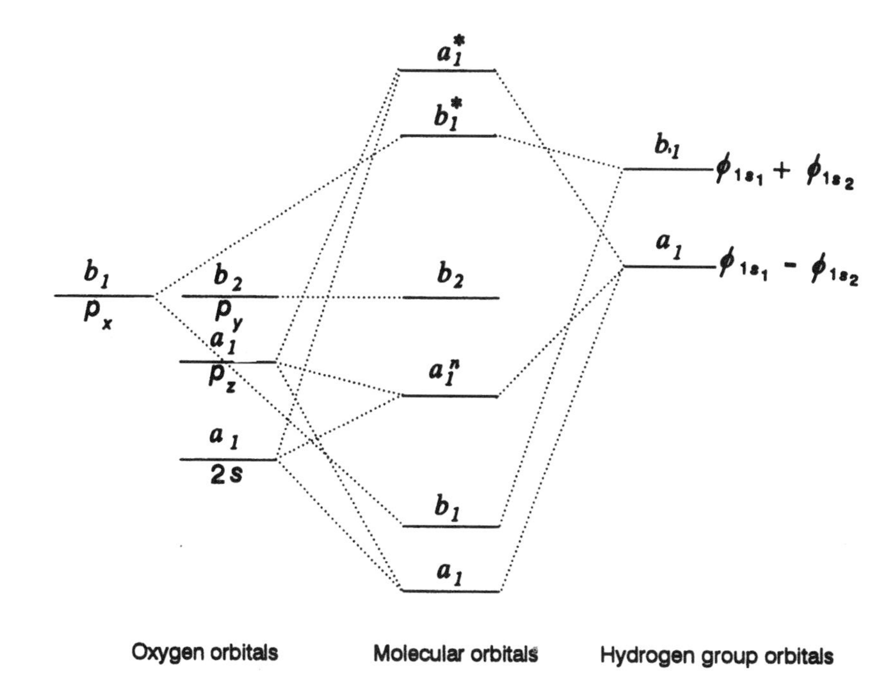

Figure 10.12 ▶ The molecular orbital diagram for the water molecule.

In doing this, we must recognize that there are *two* hydrogen $1s$ orbitals and the orbitals from the oxygen atom must interact with *both* of them. Therefore, it is not each hydrogen $1s$ orbital individually that is used, but rather a combination of the two. These combinations of atomic orbitals are called *group orbitals*, and in this case the combinations can be written as $(\phi_{1s(1)} + \phi_{1s(2)})$ and $(\phi_{1s(1)} - \phi_{1s(2)})$. The $2s$ and $2p_z$ orbitals having A_1 symmetry mix with the combination of hydrogen $1s$ orbitals, which have A_1 symmetry to produce three molecular orbitals having A_1 symmetry (one bonding, one nonbonding, and one antibonding). The $2p_x$ orbital having B_1 symmetry combines with the combination of hydrogen orbitals, which has the same symmetry, and that combination is $(\phi_{1s(1)} - \phi_{1s(2)})$. The $2p_y$ orbital, which has B_2 symmetry, remains as a π orbital and does not have the correct symmetry to interact with either of the combinations of hydrogen orbitals. In the case of the H_2O molecule, the four orbitals of lowest energy will be populated because the atoms have a total of eight valence shell electrons. Therefore, the bonding in the H_2O molecule can be represented as

$$(a_1)^2(b_1)^2(a_1^n)^2(b_2)^2.$$

As in the case of atomic orbitals and spectroscopic states (see Chapter 5), we use *lower case letters to denote orbitals or configurations* and *upper case letters to indicate states.*

Having considered the case of the C_{2v} water molecule, we would like to be able to use the same procedures to construct the qualitative molecular orbital diagrams for molecules having other structures. To do this requires that we know how the orbitals of the central atom transform when the symmetry of the molecule is not C_{2v}. Table 10.6 shows how the s and p orbitals are transformed in several common point groups, and more extensive tables can be found in the comprehensive books listed at the end of this chapter.

If we now consider a planar molecule like BF_3 (D_{3h} symmetry), the z axis is defined as the C_3 axis. One of the B–F bonds lies along the x axis as shown in Figure 10.13. The symmetry elements present for this molecule include the C_3 axis, three C_2 axes (coincident with the B–F bonds), three mirror planes each containing a C_2 axis and the C_3 axis, and the identity.

TABLE 10.6 ▶ Central Atom *s* and *p* Orbital Transformations under Different Symmetries

		Orbital			
Point group	Structure	s	p_x	p_y	p_z
C_{2v}	Bent triatomic	A_1	B_1	B_2	A_1
C_{3v}	Pyramidal	A_1	E	E	A_1
D_{3h}	Trigonal planar	A_1'	E'	E'	A_2''
C_{4v}	Pyramidal	A_1	E	E	A_1
D_{4h}	Square planar	A_1'	E_u	E_u	A_{2u}
T_d	Tetrahedral	A_1	T_2	T_2	T_2
O_h	Octahedral	A_1	T_{1u}	T_{1u}	T_{1u}
$D_{\infty h}$	Linear	Σ_g	Σ_u	Σ_u	Σ_g^+

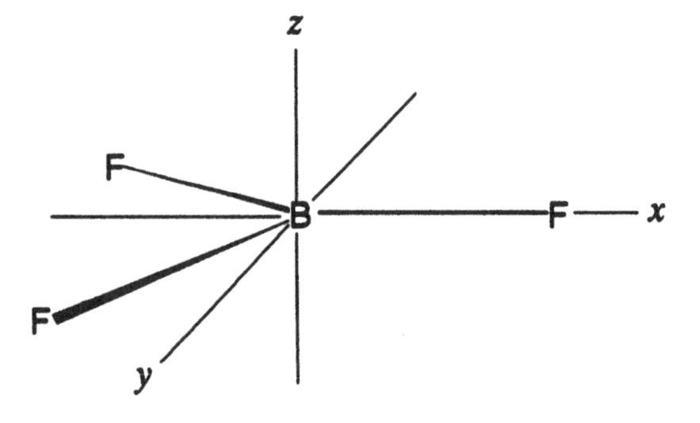

Figure 10.13 ▶ The coordinate system for BF_3.

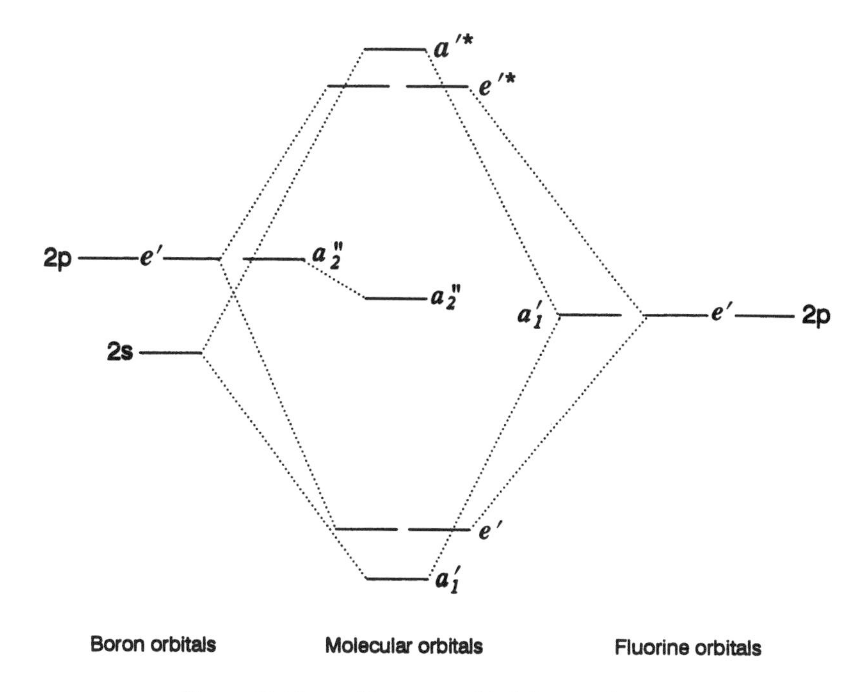

Figure 10.14 ▶ Molecular orbital diagram for BF_3.

Thus, there are 12 symmetry operations that can be performed with this molecule. It can be shown that the p_x and p_y orbitals both transform as E' and the p_z orbital transforms as A_2''. The s orbital is A_1' (the prime indicating symmetry with respect to σ_h). Similarly, we could find that the fluorine p_z orbitals are A_1, E_1, and E_1 for the three atoms. The qualitative molecular orbital diagram can then be constructed as shown in Figure 10.14. It is readily apparent that the bonding molecular orbitals (three σ bonds) are capable of holding the six bonding electrons in this molecule. The possibility of some π bonding is seen in the molecular orbital diagram due to the presence of the a_2'' orbital, and in fact there is some evidence for this type of interaction. The sum of the covalent radii of boron and fluorine atoms is about 1.52 Å, but the experimental B–F bond distance in BF_3 is about 1.295 Å. Part of this "bond shortening" may be due to partial double bonds resulting from the π bonding. Structures showing double bonds lead to three resonance structures of the valence bond type that can be shown as

From these resonance structures, we determine a bond order of 1.33, which would predict that the observed bond length should be shorter than that expected for a single bond.

Having seen the development of the molecular orbital diagrams for AB_2 and AB_3 molecules, we will now consider tetrahedral molecules like CH_4, SiH_4, or SiF_4. In this symmetry, the valence shell s orbital on the central atom transforms as A_1 while the p_x, p_y, and p_z orbitals transform as T_2 (see Table 10.6). For the methane molecule, the combination of hydrogen orbitals that transforms as A_1 is

$$\phi_{1s(1)} + \phi_{1s(2)} + \phi_{1s(3)} + \phi_{1s(4)},$$

and the combination transforming as T_2 is

$$\phi_{1s(1)} + \phi_{1s(2)} - \phi_{1s(3)} - \phi_{1s(4)},$$

where the coordinate system is as shown in Figure 10.15. Using the orbitals on the carbon atom and combining them with the group orbitals from the four hydrogen atoms (the linear combination of orbitals having a symmetry matching the carbon atom orbitals), we obtain the molecular

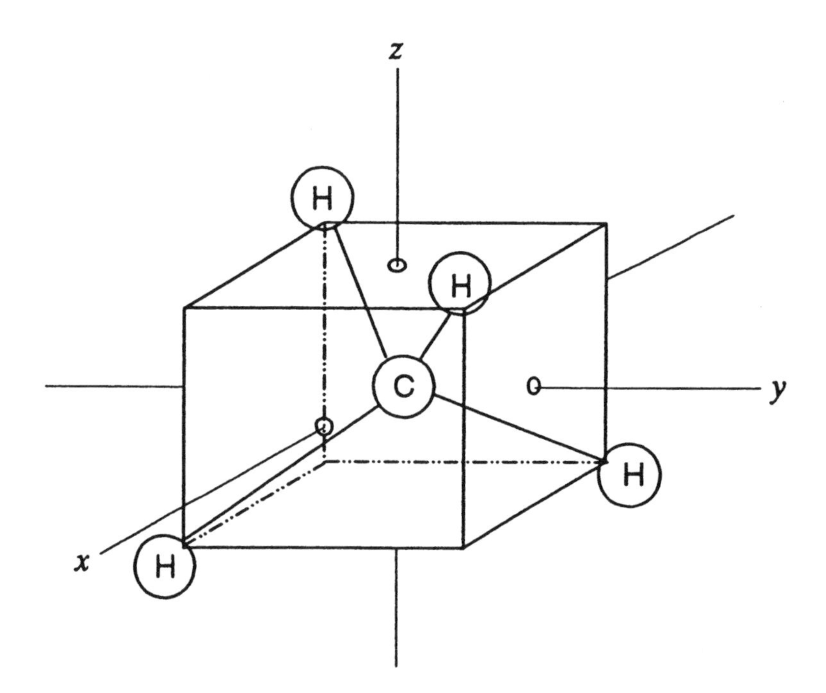

Figure 10.15 ▶ Coordinate system for the tetrahedral CH_4 molecule.

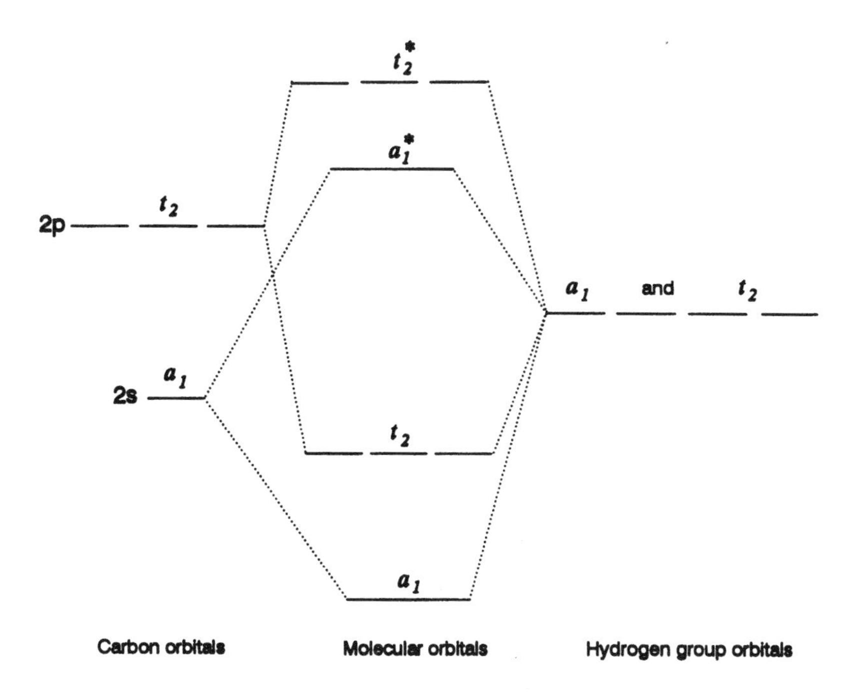

Figure 10.16 ▶ The molecular orbital diagram for CH_4.

orbital diagram shown in Figure 10.16. The hydrogen group orbitals are referred to as symmetry adjusted linear combinations (SALC) because they have symmetry that matches the orbitals of the carbon atom. The molecular orbital diagrams for other tetrahedral molecules are similar to that for CH_4.

For an octahedral AB_6 molecule such as SF_6, the valence shell orbitals of the central atom are considered to be the s, p, and d orbitals. It is easy to see that a regular octahedron has a center of symmetry so that g and u designations must be used on the symmetry species to designate symmetry or asymmetry with respect to that center. Clearly the s orbital transforms as A_{1g}. The three p orbitals, being directed toward the corners of the octahedron, are degenerate and change sign upon reflection through the center of symmetry. They thus constitute a set that can be designated as T_{1u}. Of the set of d orbitals, the d_{z^2} and $d_{x^2-y^2}$ orbitals are directed toward the corners of the octahedron and they do not change sign upon inversion through the center of symmetry. These orbitals are, therefore, designated as E_g. The remaining d_{xy}, d_{yz}, and d_{xz} orbitals form a triply degenerate set designated as T_{2g}.

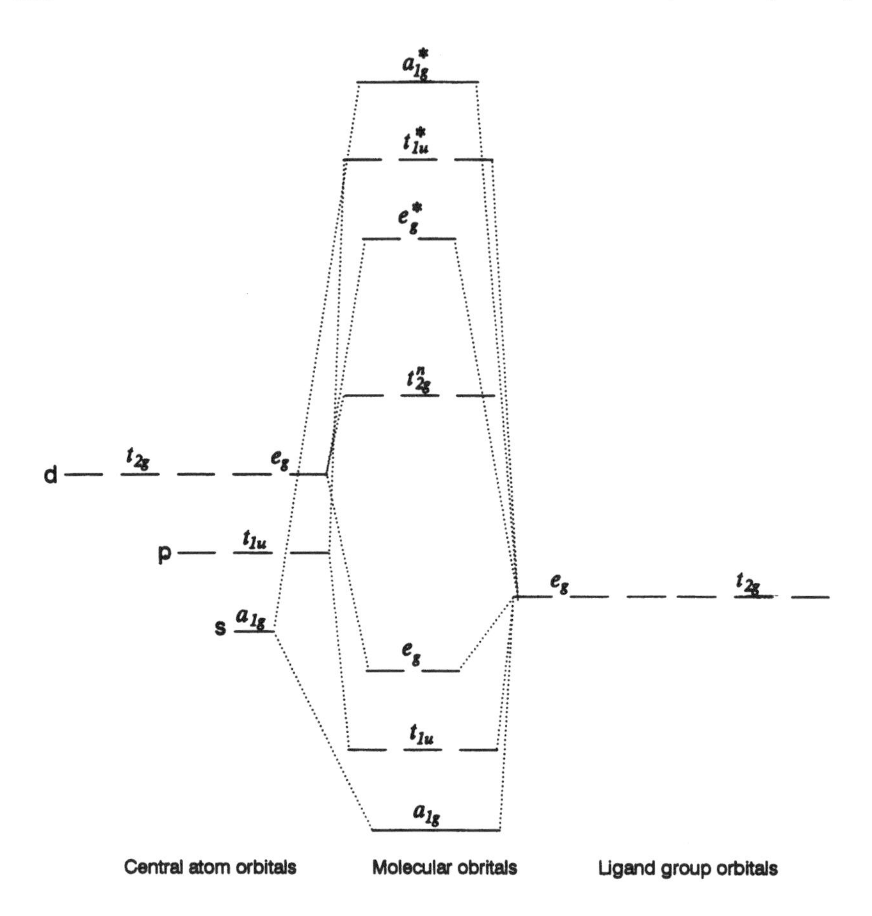

Figure 10.17 ▶ Molecular orbital diagram for an octahedral molecule.

If we consider only σ bonding, we find that T_{1u}, E_g, and A_{1g} orbitals are used in bonding to the six groups attached. The resulting energy level diagram is shown in Figure 10.17. In this section, we have seen how symmetry considerations are used to arrive at qualitative molecular orbital diagrams for molecules having several common structural types. The number of molecules and ions that have C_{2v}, C_{3v}, T_d, and O_h symmetry is indeed large. Energy level diagrams such as those shown in this section are widely used to describe structural, spectroscopic, and other properties of molecules. We have not, however, set about calculating anything. Chapter 11 presents an overview of one of the simplest types of molecular orbital calculations carried out routinely. The more sophisticated mathematical treatments of molecular orbital calculations are beyond the intended scope of this introductory book.

References for Further Reading

- Adamson, A. W. (1986). *A Textbook of Physical Chemistry*, 3rd ed., Chap. 17. Academic Press College Division, Orlando. Probably the best treatment of symmetry available in a physical chemistry text.
- Cotton, F. A. (1990). *Chemical Applications of Group Theory*, 3rd ed. Wiley, New York. The standard high-level book on group theory. Probably the most cited reference in the field.
- Drago, R. S. (1992). *Physical Methods for Chemists*, Chaps. 1 and 2. Saunders College Publishing, Philadelphia. About all the nonspecialist needs in order to use the ideas of symmetry and group theory.
- Fackler, J. P. (1971). *Symmetry in Coordination Chemistry*. Academic Press, New York. A clear introduction to symmetry.
- Harris, D. C., and Bertolucci, M. D. (1989). *Symmetry and Spectroscopy*, Chap. 1. Dover, New York. Outstanding coverage of symmetry and elementary group theory.

Problems

1. Make sketches of the listed species showing approximately correct geometry and all valence shell electrons. Identify all symmetry elements present and determine the point group for the species.

(a) OCN^-,
(b) IF_2^+,
(c) IC_4^- ,
(d) SO_3^{2-},
(e) SF_6,
(f) IF_5,
(g) ClF_3,
(h) SO_3,
(i) ClO_2^-,
(j) NSF.

2. Make sketches of the listed species showing approximately correct geometry and all valence shell electrons. Identify all symmetry elements present and determine the point group for the species.

(a) CN_2^{2-},
(b) PH_3,

(**c**) PO_3 ,

(**d**) $B_3N_3H_6$,

(**e**) SF_2,

(**f**) ClO_3,

(**g**) SF_4,

(**h**) C_3O_2,

(**i**) AlF_6^{3-},

(**j**) F_2O.

3. Consider the molecule AX_3Y_2, which has no unshared electron pairs on the central atom. Sketch the structures for all possible isomers of this compound and determine the point group to which each belongs.

4. Use the symmetry of the valence shell atomic orbitals of the central atom to construct (using appropriate hydrogen group orbitals) the molecular orbital diagrams for the following:

(**a**) BeH_2,

(**b**) HF_2^- ,

(**c**) CH_2,

(**d**) H_2S.

5. Use the symmetry of the valence shell atomic orbitals of the central atom to construct (using appropriate hydrogen group orbitals) the molecular orbital diagrams for the following:

(**a**) AlF_3,

(**b**) BH_4^-,

(**c**) SF_6,

(**d**) NF_3.

6. Consider the molecule $Cl_2B–BCl_2$.

(**a**) If the structure is planar, what is the point group of the molecule?

(**b**) Draw a structure for $Cl_2B–BCl_2$ that would have an S_4 axis.

7. Use the procedure outlined in the text to obtain the multiplication table for the C_{4v} point group.

8. Follow the procedure used in the text in obtaining the character table for the C_{2v} point group and develop the character table for the C_{3v} point group.

9. Determine all of the symmetry elements possessed by the CH_4 molecule and give the point group for this molecule. In succession, replace hydrogen atoms with fluorine, chlorine, and bromine and determine what symmetry elements are present for each product. Determine the point group to which each product belongs.

Chapter 11

Hückel Molecular Orbital Methods

When the phrase "molecular orbital calculations" is first encountered, the mental image may well be one of hopelessly complicated mathematics and piles upon piles of computer output. It is interesting to note, however, that sometimes a relatively simple calculation may provide useful information that correlates well with experimental observations. Such is the case with the method known as the Hückel molecular orbital (HMO theory) calculation. This method was developed in 1931 by Erich Hückel, a physicist in Marburg, Germany, who was trying to understand the concept of aromaticity in benzene. The calculational procedures are relatively simple and have become known as the "back of an envelope" calculations. The extended Hückel method (EHMO) will be described briefly and the more sophisticated methods for dealing with the structure of molecules will be treated in Chapter 12.

11.1 The Hückel Method

In the Hückel method, the assumption is made that the σ and π parts of the bonding in molecules can be separated. Also, the overlap of orbitals on nonadjacent atoms is assumed to be zero. Additionally, the interaction energy between atoms not adjacent is assumed to be zero. The energies to be considered are represented as H_{ii}, the valence state ionization potential

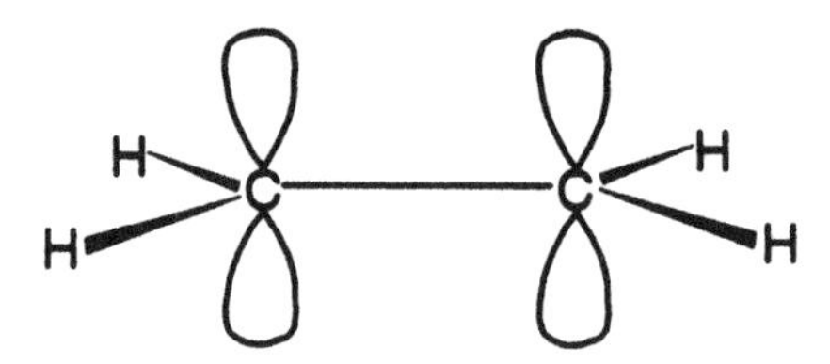

Figure 11.1 ▶ The ethylene molecule showing the p orbitals used in π bonding.

(VSIP) for atom i, and H_{ij}, the exchange energy between atoms i and j. In the Hückel treatment, it is assumed that $H_{ij} = 0$ when $|i - j| \geq 2$ (nonadjacent atoms).

When the overlap is not included in the calculations, we let $S_{12} = S_{21} = 0$. Let us illustrate the Hückel method by treating the π bond in the C_2H_4 molecule, which is shown in Figure 11.1. In this case, we will consider the carbon atoms to use sp^2 hybrid orbitals to form the σ bonds, and the p_z orbitals (which are perpendicular to the plane of the molecule) are left to form the π bonds. The wave functions for the bonding orbitals can be written as

$$\psi\text{CH}(\sigma) = a_1\phi(1s) + a_2\phi(sp^2) \tag{11.1}$$

$$\psi\text{CC}(\sigma) = a_1\phi(sp^2) + a_2\phi(sp^2) \tag{11.2}$$

$$\psi\text{CC}(\pi) = a_1\phi(p_1) + a_2\phi(p_2), \tag{11.3}$$

where $\psi\text{CC}(\sigma)$ corresponds to the σ bond between two carbon atoms, etc. Using the procedure developed earlier (see Chapter 9), we can write the secular determinant as

$$\begin{vmatrix} H_{11} - E & H_{12} - S_{12}E \\ H_{21} - S_{21} & H_{22} - E \end{vmatrix} = 0. \tag{11.4}$$

The parameters representing energies are denoted as

$$\alpha = H_{11} = H_{22} = H_{33} = \cdots = H_{ii} = H_{jj} = \text{the Coulomb integral and}$$
$$\beta = H_{12} = H_{21} = H_{23} = H_{32} = \cdots = H_{ij} = \text{the Exchange integral.}$$

Therefore, the secular determinant can be written as

$$\begin{vmatrix} \alpha - E & \beta \\ \beta & \alpha - E \end{vmatrix} = 0. \tag{11.5}$$

If we divide each element in the determinant by β, we obtain

$$\begin{vmatrix} \frac{\alpha-E}{\beta} & 1 \\ 1 & \frac{\alpha-E}{\beta} \end{vmatrix} = 0. \tag{11.6}$$

If we now let $x = (a - E)/\beta$, the determinant can be written as

$$\begin{vmatrix} x & 1 \\ 1 & x \end{vmatrix} = 0. \tag{11.7}$$

Expansion of this determinant gives

$$x^2 - 1 = 0. \tag{11.8}$$

The roots of this equation are $x = -1$ and $x = +1$. Therefore, we can write

$$\frac{\alpha - E}{\beta} = -1 \quad \text{so that } E = \alpha + \beta \tag{11.9}$$

$$\frac{\alpha - E}{\beta} = 1 \quad \text{so that } E = \alpha - \beta. \tag{11.10}$$

When we recall that both α and β are negative energies, we see that $E = \alpha + \beta$ corresponds to the state having lower energy. An energy level diagram showing the states can be constructed as shown in Figure 11.2. Each carbon atom contributes one electron to the π bond so that the lowest level is filled with two electrons. If the two electrons were on isolated carbon atoms, their binding energies would be 2α. Therefore, the energy for the two electrons in the *molecular* orbital, as opposed to what the energy would be for two isolated atoms, is

$$2(\alpha + \beta) - 2\alpha = 2\beta. \tag{11.11}$$

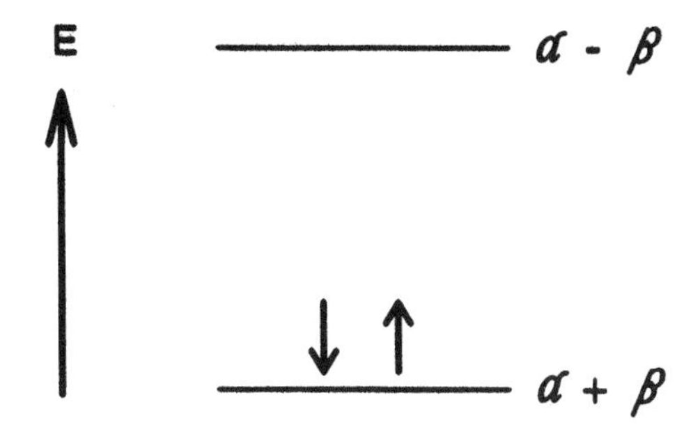

Figure 11.2 ▶ Energy level diagram for ethylene.

The fact that the molecular orbital encompasses the complete carbon structure while the atomic orbitals do not leads to the description of the molecular orbital as being *delocalized.* The energy 2β is called the *delocalization* or *resonance energy*. An average energy for a typical C–C bond is about 335–355 kJ/mol, while that for a typical C=C bond is about 600 kJ/mol. Therefore, the π bond adds about 250 kJ/mol, which means that this is the approximate value of 2β, so that β is about 125 kJ/mol.

We have obtained the energies of the two molecular orbitals (bonding and antibonding), but a major objective of the calculations is to be able to obtain useful information from the wave functions. In order to do this, we need to determine the values of the constants a_1 and a_2. These constants determine the weighting given to the atomic wave functions and thereby determine electron density, etc. The wave function for the bonding orbital in ethylene can be written as

$$\psi_b = a_1\phi_1 + a_2\phi_2, \tag{11.12}$$

and we know that for the normalized wave function,

$$\int \psi_b^2 \, d\tau = 1 = \int (a_1\phi_1 + a_2\phi_2)^2 \, d\tau. \tag{11.13}$$

Expansion of the integral and letting S_{11}, S_{22}, and S_{12} represent the overlap integrals we obtain

$$a_1^2 S_{11} + a_2^2 S_{22} + 2a_1a_2S_{12} = 1. \tag{11.14}$$

If we let $S_{11} = S_{22} = 1$ and neglect the overlap between adjacent atoms by letting $S_{12} = S_{21} = 0$, then

$$a_1^2 + a_2^2 = 1. \tag{11.15}$$

The secular equations can be written in terms of α and β as

$$a_1(\alpha - E) + a_2\beta = 0 \tag{11.16}$$

$$a_1\beta + a_2(\alpha - E) = 0. \tag{11.17}$$

Dividing both equations by β and letting $x = (a - E)/\beta$, we find that

$$a_1x + a_2 = 0 \tag{11.18}$$

$$a_1 + a_2x = 0. \tag{11.19}$$

For the bonding state, we saw that $x = -1$, so it follows that $a_1^2 = a_2^2$. Therefore, we can write

$$a_1^2 + a_1^2 = 1 = 2a_1^2 \tag{11.20}$$

so that $a_1 = 1/(2)^{1/2} = a_2$ and the wave function for the bonding state can be written in terms of the atomic wave functions as

$$\psi_b = 0.707\phi_1 + 0.707\phi_2. \tag{11.21}$$

For an atom in a molecule, it is the square of the coefficient of the atomic wave function in the molecular wave function that gives the probability (density) of finding an electron on that atom. Therefore, $a_1^2 = a_2^2 = \frac{1}{2}$, so $\frac{1}{2}$ of the electrons should be on each atom. Since there are two electrons in the bonding orbital, the electron density (ED) is $2(\frac{1}{2}) = 1$ and one electron resides on each atom. As expected electrons are not transferred from one carbon atom to the other.

Another useful property for describing bonding in the molecule is the bond order, B. This quantity gives an electron population in terms of the number of π bonds between two bonded atoms. In this case, it is the product of the coefficients on the atomic wave functions that gives the density of the bond between them. However, we must also take into account the total number of electrons in the occupied molecular orbital(s). Therefore, the bond order between atoms X and Y can be written as B_{XY}, which is given by

$$B_{XY} = \sum_{i=1}^{n} a_X a_Y p_i, \tag{11.22}$$

where a is the weighting coefficient, n is the number of populated orbitals, and p is the population (number of electrons) in that orbital. For the case of ethylene, there is only one orbital and it is populated with two electrons, so we find that

$$B_{CC} = 2(0.707)(0.707) = 1, \tag{11.23}$$

which indicates that there is one π bond between the carbon atoms. Before considering larger molecules we will show some of the mathematics necessary for application of the Hückel method.

11.2 Determinants

From the example in the previous section, it should be apparent that formulating a problem using the HMO method to describe a molecule results in a *determinant*. A determinant represents a function in the form of an array that contains *elements* in *rows* and *columns*. We also encountered a secular determinant in dealing with diatomic molecules earlier (see Section 9.3).

The number of rows or columns (they are equal) is called the rank (or *order*) of the determinant. Determinants are essential in the Hückel method so we need to show how they are manipulated.

A determinant can be reduced to an equation known as the *characteristic equation*. Suppose we consider the 2×2 determinant

$$\begin{vmatrix} a & b \\ c & d \end{vmatrix} = 0. \tag{11.24}$$

Simplifying this determinant to obtain the characteristic equation involves multiplying along one diagonal and subtracting the product obtained by multiplying along the other diagonal. This rule applied to the preceding determinant gives

$$\begin{vmatrix} a & b \\ c & d \end{vmatrix} = ad - bc = 0. \tag{11.25}$$

If a determinant can be written as

$$\begin{vmatrix} x & 1 \\ 1 & x \end{vmatrix} = 0 \tag{11.26}$$

we obtain the characteristic equation

$$x^2 - 1 = 0, \tag{11.27}$$

which is the equation that arises from the treatment of the π bond in ethylene (see Section 11.1).

If the molecule being considered contains three atoms, we will obtain a 3×3 determinant such as

$$\begin{vmatrix} a & b & c \\ d & e & f \\ g & h & i \end{vmatrix} = 0. \tag{11.28}$$

Expansion of a 3×3 determinant is somewhat more elaborate than that of a 2×2 determinant. One method can be illustrated as follows: Initially, extend the determinant by writing the first two columns again to the right of the determinant.

$$\left|\begin{matrix} a & b & c \\ d & e & f \\ g & h & i \end{matrix}\right| \begin{matrix} a & b \\ d & e \\ g & h \end{matrix}.$$

We now perform the multiplication along each three-membered diagonals, with multiplication to the *right* giving products that are *positive* and multiplication to the *left* giving products that are *negative*. Therefore, in this case, we obtain the characteristic equation

$$aei + bfg + cdh - ceg - afh - bdi = 0. \tag{11.29}$$

If the determinant being considered is

$$\begin{vmatrix} x & 1 & 0 \\ 1 & x & 1 \\ 0 & 1 & x \end{vmatrix} = 0, \tag{11.30}$$

we expand the determinant as follows:

$$\begin{vmatrix} x & 1 & 0 \\ 1 & x & 1 \\ 0 & 1 & x \end{vmatrix} \begin{matrix} x & 1 \\ 1 & x \\ 0 & 1 \end{matrix}$$

This method results in the characteristic equation that reduces to

$$x^3 - 2x = 0, \tag{11.31}$$

and the roots are $x = 0, -(2)^{1/2}$, and $+(2)^{1/2}$. However, this expansion method using diagonals works only for 2×2 and 3×3 determinants, so we need a more general method for expansion of higher order determinants. Consider the 4×4 determinant

$$\begin{vmatrix} a & b & c & d \\ e & f & g & h \\ i & j & k & l \\ m & n & o & p \end{vmatrix} = 0. \tag{11.32}$$

Expansion of this determinant is accomplished by a procedure known as the *method of cofactors*. In this method, we begin with element a and remove the row and column that contain a. Then the rest of the determinant is multiplied by a to obtain

$$a \begin{vmatrix} f & g & h \\ j & k & l \\ n & o & p \end{vmatrix}.$$

The portion of the determinant multiplied by a is called a *minor* and will have a rank $(n - 1)$, where n is the rank of the original determinant. We now repeat this process except that b is removed and the sign preceding this term is negative, and so on. Therefore,

$$\begin{vmatrix} a & b & c & d \\ e & f & g & h \\ i & j & k & l \\ m & n & o & p \end{vmatrix} = a\begin{vmatrix} f & g & h \\ j & k & l \\ n & o & p \end{vmatrix} - b\begin{vmatrix} e & g & h \\ i & k & l \\ m & o & p \end{vmatrix} + c\begin{vmatrix} e & f & h \\ i & j & l \\ m & n & p \end{vmatrix} - d\begin{vmatrix} e & f & g \\ i & j & k \\ m & n & o \end{vmatrix}. \tag{11.33}$$

We can now continue the expansion of each 3×3 determinant as previously illustrated.

The cofactors are designated as C_{ij} and their general formula is

$$C_{ij} = (-1)^{i+j} M_{ij}, \tag{11.34}$$

where M_{ij} is the minor having rank $(n - 1)$. For the determinant

$$\begin{vmatrix} x & 1 & 0 & 0 \\ 1 & x & 1 & 0 \\ 0 & 1 & x & 1 \\ 0 & 0 & 1 & x \end{vmatrix} = 0, \tag{11.35}$$

we can show the expansion to find the characteristic equation as

$$\begin{vmatrix} x & 1 & 0 & 0 \\ 1 & x & 1 & 0 \\ 0 & 1 & x & 1 \\ 0 & 0 & 1 & x \end{vmatrix} = x\begin{vmatrix} x & 1 & 0 \\ 1 & x & 1 \\ 0 & 1 & x \end{vmatrix} - 1\begin{vmatrix} 1 & 1 & 0 \\ 0 & x & 1 \\ 0 & 1 & x \end{vmatrix} + 0\begin{vmatrix} 1 & x & 0 \\ 0 & 1 & 1 \\ 0 & 0 & x \end{vmatrix} - 0\begin{vmatrix} 1 & x & 1 \\ 0 & 1 & x \\ 0 & 0 & 1 \end{vmatrix}. \tag{11.36}$$

Since the last two terms give 0, the characteristic equation can be written as

$$x^4 - 3x^2 + 1 = 0. \tag{11.37}$$

Another useful property of determinants is illustrated by *Laplace's Expansion Theorem*, which relates to the expansion of a determinant in terms of smaller units or subdeterminants. Sometimes, because of the symmetry of the determinant, it is possible to simplify the determinant as is illustrated by the following example. Consider the determinant

$$\begin{vmatrix} x^2 & 4 & 0 & 0 \\ x & 3 & 0 & 0 \\ 0 & 0 & 12 & 2x \\ 0 & 0 & 5x & 2 \end{vmatrix} = f(x). \tag{11.38}$$

In this case, the function represented can be written as

$$\begin{vmatrix} x^2 & 4 \\ x & 3 \end{vmatrix} \cdot \begin{vmatrix} 12 & 2x \\ 5x & 2 \end{vmatrix} = f(x). \tag{11.39}$$

Expanding each of the 2×2 determinants is carried out as illustrated earlier to give the characteristic equation

$$\begin{aligned} f(x) &= \left(3x^2 - 4x\right)\left(24 - 10x^2\right) \\ f(x) &= -30x^4 + 40x^3 + 72x^2 - 96x. \end{aligned} \tag{11.40}$$

This technique is useful in simplifying the secular determinants that arise for some organic molecules having certain structures. Because only interactions between adjacent atoms are considered, several elements in the secular determinant can be set equal to 0.

Determinants also have other useful and interesting properties that will simply be listed here. For a complete discussion of the mathematics of determinants, see the reference books listed at the end of this chapter.

1. Interchanging two rows (or columns) of a determinant produces a determinant that is the negative of the original.

2. If each element in a row (or column) of a determinant is multiplied by a constant, the result is the constant times the original determinant. If we consider the determinant

$$D = \begin{vmatrix} a & b \\ c & d \end{vmatrix}, \tag{11.41}$$

we can now multiply each member of the second column by k,

$$D = \begin{vmatrix} a & kb \\ c & kd \end{vmatrix} \tag{11.42}$$

and expansion gives

$$adk - bck = k\,(ad - bc) = k\begin{vmatrix} a & b \\ c & d \end{vmatrix}.$$

3. If two rows (or columns) in a determinant are identical, the determinant has the value of 0. For example,

$$\begin{vmatrix} a & a & b \\ c & c & d \\ e & e & f \end{vmatrix} = acf + ade + bce - bce - ade - acf = 0. \quad (11.43)$$

4. If a determinant has one row (or column) where each element is zero, the determinant evaluates to 0. This is very easy to verify so an example will not be provided here.

As we shall see later in this chapter, evaluation of determinants is an integral part of the HMO method.

11.3 Solving Polynomial Equations

We have just seen that the expansion of determinants leads to polynomial equations that we will need to solve. Before the widespread availability of sophisticated calculators, such equations were solved graphically. In fact, a very popular book in the early 1960s was John D. Roberts' *Notes on Molecular Orbital Calculations* (Benjamin, 1962). That book popularized the application of the Hückel method in organic chemistry.

Roots of the polynomial equations can often be found by graphing the functions if they cannot be factored directly. As we shall see, the equation

$$x^4 - 3x^2 + 1 = 0 \quad (11.44)$$

arises from the expansion of the secular determinant for butadiene. To find the values of x where this function $y = f(x)$ crosses the x axis, we let

$$y = x^4 - 3x^2 + 1 = 0 \quad (11.45)$$

and make the graph by assigning values to x. We can get a rough estimate of the range of x values to try by realizing that for large values of x, x^4 and

$3x^2$ are both much greater than 1. Therefore, for y to have a value near 0, the terms x^4 and $3x^2$ must be approximately equal,

$$x^4 \approx 3x^2$$
$$x^2 \approx 3$$
$$x \approx \pm\sqrt{3}.$$

Therefore, we choose the range of values $2 > x > -2$, which should encompass the roots. The values of y when x is assigned values in this range are shown below.

x	y	x	y
−2.0	5.0	0.4	0.546
−1.6	−0.126	0.8	−0.510
−1.2	−1.254	1.2	−1.254
−0.8	−0.510	1.6	−0.126
−0.4	0.546	2.0	5.0
0.0	1.000		

The graph of these values is shown in Figure 11.3. It is apparent from the data or the graph that the four roots are approximately ±0.6 and ±1.6. In this case, the actual roots of Eq. (11.45) are

$$x = -1.62, \quad -0.618, \quad +0.618, \text{ and } +1.62.$$

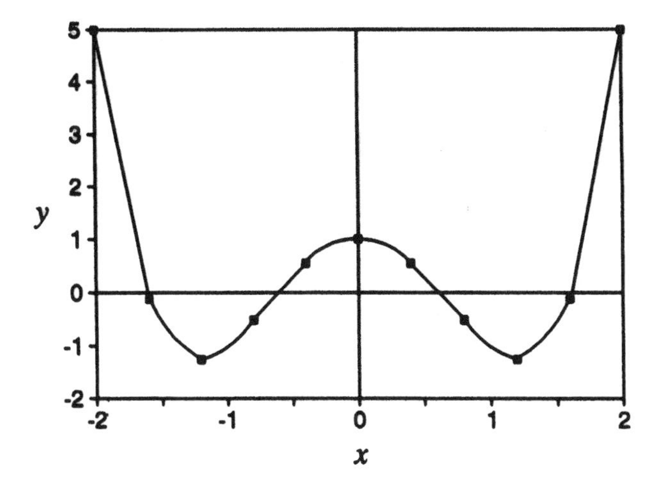

Figure 11.3 ▶ A graph of the data shown in the text for $y = x^4 - 3x^2 + 1$.

The advent of the graphing calculator has made this procedure very rapid to carry out. At most schools, calculus is now taught in such a way as to require students to use these electronic marvels. Solving the polynomial equations resulting from Hückel MO calculations is indeed very easy. Also, many calculators that are not graphing ones have a built in "solve" function. This lets the user enter the expression to be solved and obtain the roots of polynomial equations in a convenient manner. Most of these "hard-wired" capabilities make use of the *Newton–Raphson* or *secant* methods that are standard techniques in numerical analysis. For details of how these methods work, see the references on numerical analysis cited at the end of the chapter. Electronic calculators and computer software has made possible the routine use of sophisticated numerical analysis procedures.

Another numerical technique useful in certain types of problems is *iteration*. In an iterative technique, some operation is used repetitively in order to solve a problem. A programmable calculator or computer is ideally suited to this type of calculation. Finding the roots of polynomial equations like those arising from Hückel calculations is essentially finding the zeros of functions (values of x where $y = 0$). The roots are the values of x that satisfy the equation

$$a_n x^n + a_{n-1} x^{n-1} + \cdots + a_1 x + a_0 = 0. \tag{11.46}$$

From the quadratic formula, an equation that can be written

$$ax^2 + bx + c = 0 \tag{11.47}$$

has the roots

$$x = \frac{-b \pm \sqrt{b^2 - 4ac}}{2a}. \tag{11.48}$$

If $b^2 - 4ac$ (called the *discriminant*) is negative, the roots of the equation are complex. If $b^2 - 4ac$ is positive, the roots are real. Most polynomial equations arising from secular determinants will be higher than second order so an analytical method for their solution is not nearly as simple as it is for a quadratic equation. Suppose we wish to solve the equation

$$x^2 + 3.6x - 16.4 = 0. \tag{11.49}$$

Applying the quadratic formula we obtain

$$x = \frac{-3.6 \pm \sqrt{3.6^2 - 4(-16.4)}}{2} = +2.63, -6.23. \tag{11.50}$$

After writing Eq. (11.49) as $x(x + 3.6) = 16.4$, we can write

$$x = \frac{16.4}{x + 3.6}. \tag{11.51}$$

From the equation written in this form we see that if we let $x = 2$, we would find that

$$2 = \frac{16.4}{2 + 3.6} = 2.28 \quad \text{(which is > 2)}.$$

Therefore, 2 is not a correct answer, but one root must lie between 2 and 3 because if we let $x = 3$,

$$3 = \frac{16.4}{3 + 3.6} = 1.52 \quad \text{(which is < 2)}.$$

Suppose we call $x = 2$ a "first guess" at a solution, let it equal x_0. The next value, x_1, can be written in terms of x_0 as

$$x_1 = \frac{16.4}{x_0 + 3.6} = \frac{16.4}{2 + 3.6} = 2.28. \tag{11.52}$$

Now, we will use this value as an "improved" guess and calculate a new value, x_2.

$$x_2 = \frac{16.4}{x_1 + 3.6} = \frac{16.4}{2.28 + 3.6} = 2.79.$$

Repeating the process with each new value of x_{n+1}, being given in terms of x_n,

$$x_3 = \frac{16.4}{x_2 + 3.6} = \frac{16.4}{2.79 + 3.6} = 2.57$$

$$x_4 = \frac{16.4}{x_3 + 3.6} = \frac{16.4}{2.57 + 3.6} = 2.66$$

$$x_5 = \frac{16.4}{x_4 + 3.6} = \frac{16.4}{2.66 + 3.6} = 2.62, \text{ etc.}$$

In this process, we are using the formula (known as a recursion formula)

$$x_{n+1} = \frac{16.4}{x_n + 3.6}. \tag{11.53}$$

As we saw from Eq. (11.50), one root is $x = 2.63$, so it is apparent that this iterative process is converging to a correct root.

If we write Eq. (11.53) in a general form as

$$x = f(x), \tag{11.54}$$

it can be shown that the iterative process will converge if $\left|f'(x)\right| < 1$. In the preceding case,

$$f(x) = \frac{16.4}{x + 3.6}, \tag{11.55}$$

and taking the derivative gives

$$f'(x) = \frac{-16.4}{(x + 3.6)^2}. \tag{11.56}$$

For the root where $x = 2.63$, $\left|f'(x)\right| = 0.423$, which is less than 1 and convergence is achieved. In the general case using the equation $x^2 + bx + c = 0$,

$$x = \frac{-c}{x + b} \tag{11.57}$$

and

$$f'(x) = \frac{c}{(x + b)^2}. \tag{11.58}$$

If $\left|(x + b)^2\right| \approx c$, convergence will be slow. Verify that this is indeed true using the equation

$$x^2 - 4x + 3.99 = 0. \tag{11.59}$$

For an equation like

$$x^2 - 5.0x + 7.5 = 0, \tag{11.60}$$

the roots are complex because $b^2 - 4ac$ is negative, and the iterative method does not work.

Iterative methods are well suited to solving equations like

$$x - (1 + \cos x)^{1/2} = 0. \tag{11.61}$$

It is easy to see that for *some* value of x,

$$x = (1 + \cos x)^{1/2} \quad \text{(when } x \text{ is in radians)}. \tag{11.62}$$

A graph showing $y = f(x)$ can now be prepared for each of the two sides of the equation (Figure 11.4). From the graphs we can see that x is

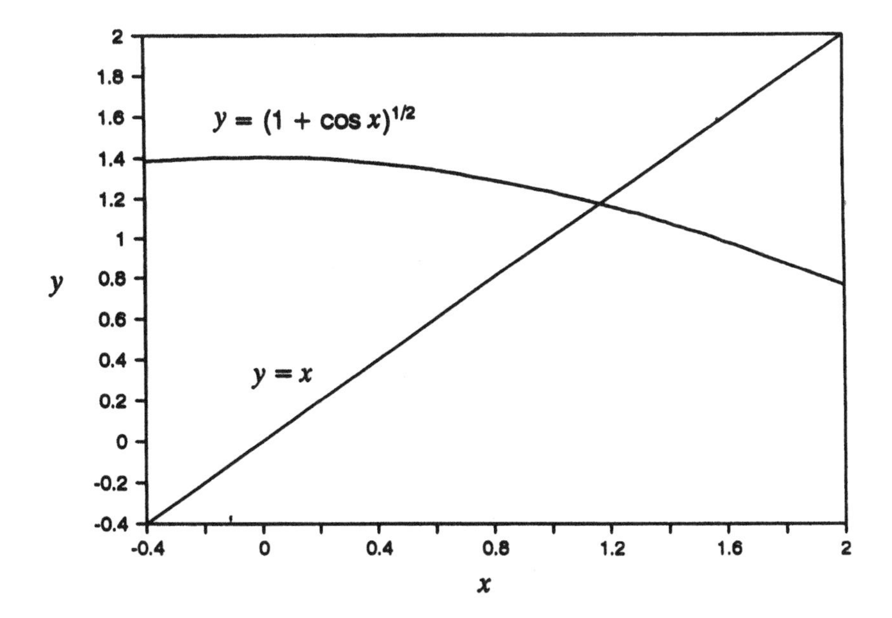

Figure 11.4 ▶ The graphs showing $y = x$ and $y = (1 + \cos x)^{1/2}$.

approximately 1, so we can begin an iterative process using a trial value of $x_0 = 1$. Then

$$
\begin{aligned}
x_1 &= (1 + \cos 1)^{1/2} = 1.241 \\
x_2 &= (1 + \cos 1.241)^{1/2} = 1.151 \\
x_3 &= (1 + \cos 1.151)^{1/2} = 1.187 \\
x_4 &= (1 + \cos 1.187)^{1/2} = 1.173 \\
x_5 &= (1 + \cos 1.173)^{1/2} = 1.178 \\
x_6 &= (1 + \cos 1.178)^{1/2} = 1.176 \\
x_7 &= (1 + \cos 1.176)^{1/2} = 1.177 \\
x_8 &= (1 + \cos 1.177)^{1/2} = 1.1764.
\end{aligned}
$$

We have found that the root of Eq. (11.61) is 1.176.

Sometimes, problems in the sciences and engineering lead to some very interesting equations to be solved. Although such transcendental equations as those just solved do not arise very often, it is useful to know that a procedure for solving them easily exists.

11.4 Hückel Calculations for Larger Molecules

The system containing three carbon atoms in a chain shown in Figure 11.5 is that of the allyl species, which includes the neutral radical as well as the carbocation and the anion. The combination of atomic wave functions will be constructed using $2p$ wave functions from the carbon atoms. In this case, the Coulomb integrals will be equal so $H_{11} = H_{22} = H_{33}$, and the exchange integrals between adjacent atoms will be set equal, $H_{12} = H_{21} = H_{23} = H_{32}$. However, in this approximate method, interaction between atoms that are not adjacent is ignored. Thus, $H_{13} = H_{31} = 0$. All overlap integrals of the type S_{ii} are set equal to 1 and all overlap between adjacent atoms is neglected ($S_{ij} = 0$).

The secular determinant is written as follows when the usual substitutions are made:

$$\begin{vmatrix} H_{11}-E & H_{12} & 0 \\ H_{21} & H_{22}-E & H_{23} \\ 0 & H_{32} & H_{33}-E \end{vmatrix} = \begin{vmatrix} \alpha-E & \beta & 0 \\ \beta & \alpha-E & \beta \\ 0 & \beta & \alpha-E \end{vmatrix} \tag{11.63}$$

$$= \begin{vmatrix} x & 1 & 0 \\ 1 & x & 1 \\ 0 & 1 & x \end{vmatrix}.$$

By the techniques shown in Section 11.2, the characteristic equation can be written as

$$x^3 - 2x = 0, \tag{11.64}$$

which has the roots $x = 0, -(2)^{1/2}$, and $(2)^{1/2}$. Therefore, we obtain the energies of the molecular orbitals from these roots as

$$\frac{\alpha - E}{\beta} = -\sqrt{2} \qquad \frac{\alpha - E}{\beta} = 0 \qquad \frac{\alpha - E}{\beta} = \sqrt{2}$$
$$E = \alpha + \sqrt{2}\beta \qquad E = \alpha \qquad E = \alpha - \sqrt{2}\beta.$$

Both α and β are negative quantities so the lowest energy is $E = a + (2)^{1/2}\beta$. The molecular orbital diagrams can be shown as in Figure 11.6.

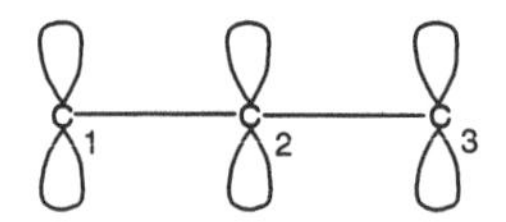

Figure 11.5 ▶ The allyl model showing the p orbitals used in π bonding.

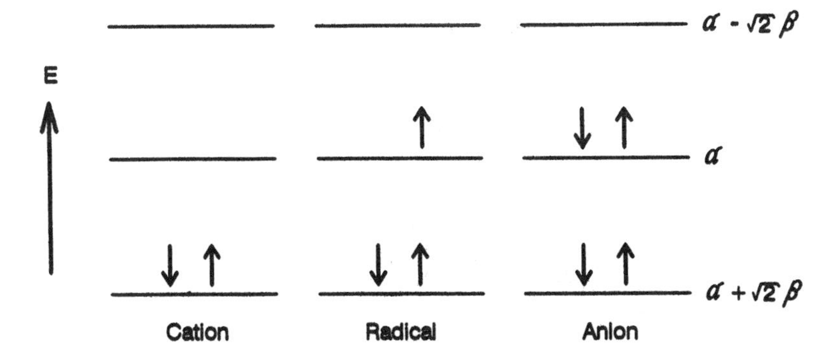

Figure 11.6 ▸ Energy level diagram for the allyl cation, radical, and anion.

The electrons have been placed in the orbitals to show the radical, the cation, and the anion.

We now proceed to evaluate the coefficients, the electron densities, and the bond orders. The elements in the secular determinant represent the coefficients in secular equations. Therefore, we can write directly

$$\begin{vmatrix} \alpha_1 x & a_2 & 0 \\ a_1 & a_2 x & a_3 \\ 0 & a_2 & a_3 x \end{vmatrix} = 0. \tag{11.65}$$

From Eq. (11.65) we see that the three equations are represented as

$$a_1 x + a_2 = 0 \tag{11.66}$$

$$a_1 + a_2 x + a_3 = 0 \tag{11.67}$$

$$a_2 + a_3 x = 0. \tag{11.68}$$

Taking first the root $x = -(2)^{1/2}$, which corresponds to the molecular orbital of lowest energy, substitution of that value in the equations above gives

$$-a_1\sqrt{2} + a_2 = 0 \tag{11.69}$$

$$a_1 - a_2\sqrt{2} + a_3 = 0 \tag{11.70}$$

$$a_2 - a_3\sqrt{2} = 0. \tag{11.71}$$

From Eq. (11.69) we find that $a_2 = (2)^{1/2} a_1$, and from Eq. (11.71) we find that $a_2 = (2)^{1/2} a_3$. Therefore, it is clear that $a_1 = a_3$, and substituting this value in Eq. (11.70) gives

$$a_1 - a_2\sqrt{2} + a_1 = 0. \tag{11.72}$$

Solving this equation for a_2, we find that $a_2 = (2)^{1/2}a_1$. We now make use of the normalization requirement that

$$a_1^2 + a_2^2 + a_3^2 = 1, \tag{11.73}$$

and by substitution for a_1,

$$a_1^2 + 2a_1^2 + a_1^2 = 4a_1^2 = 1. \tag{11.74}$$

We find that $a_1 = 0.5 = a_3$ and $a_2 = (2)^{1/2}/2 = 0.707$.

Using these coefficients for the atomic orbitals in the expression for the molecular wave function gives

$$\psi = 0.500\phi_1 + 0.707\phi_2 + 0.500\phi_3. \tag{11.75}$$

Next we use the root $x = 0$ with the secular equations, which leads to

$$0a_1 + a_2 + 0 = 0 \tag{11.76}$$

$$a_1 + 0a_2 + a_3 = 0 \tag{11.77}$$

$$0 + a_2 + 0a_3 = 0. \tag{11.78}$$

From these equations it is easy to verify that $a_2 = 0$ and $a_1 + a_3 = 0$, so $a_1 = -a_3$. Therefore, the requirement that the sum of the squares of the coefficients is equal to 1 gives

$$a_1^2 + a_3^2 = 1 = 2a_1^2, \tag{11.79}$$

from which we find $a_1 = 1/(2)^{1/2}$ and $a_3 = -1/(2)^{1/2}$. These coefficients obtained from the $x = 0$ root lead to the wave function

$$\psi_2 = 0.707\phi_1 - 0.707\phi_3. \tag{11.80}$$

When the root $x = (2)^{1/2}$ is used to evaluate the constants by the procedures above, the wave function obtained can be written as

$$\psi_3 = 0.500\phi_1 - 0.707\phi_2 + 0.500\phi_3. \tag{11.81}$$

However, this orbital remains unpopulated regardless of whether the radical, the cation, or the anion is considered (see Figure 11.6).

We can now calculate the electron densities and bond orders in the cation, radical, and anion species. For the allyl radical, there are three electrons in the π system with only one electron in the state having $E = \alpha$. Since the square of the coefficients of the atomic wave functions multiplied by the occupancy of the orbitals gives the electron density (ED), we see that

$$\begin{aligned}
\text{ED at } C_1 &= 2(0.500)^2 + 1(0.707)^2 = 1.00 \\
\text{ED at } C_2 &= 2(0.707)^2 + 1(0)^2 = 1.00 \\
\text{ED at } C_3 &= 2(0.500)^2 + 1(-0.707)^2 = 1.00
\end{aligned}$$

As expected, the three electrons are distributed equally on the three carbon atoms. The allyl carbocation has only the orbital of lowest energy ($E = \alpha + \sqrt{2}\beta$) populated with two electrons so the electron densities are

$$\begin{aligned}
\text{ED at } C_1 &= 2(0.500)^2 = 0.50 \\
\text{ED at } C_2 &= 2(0.707)^2 = 1.00 \\
\text{ED at } C_3 &= 2(0.500)^2 = 0.50.
\end{aligned}$$

For the anion, the two molecular orbitals of lowest energy are occupied with two electrons in each and the electron densities are

$$\begin{aligned}
\text{ED at } C_1 &= 2(0.500)^2 + 2(0.707)^2 = 1.50 \\
\text{ED at } C_2 &= 2(0.707)^2 + 2(0)^2 = 1.00 \\
\text{ED at } C_3 &= 2(0.500)^2 + 2(-0.707)^2 = 1.50.
\end{aligned}$$

From these results, we see that the additional electron that the anion contains compared to the radical resides in a molecular orbital centered on the terminal carbon atoms. It is to be expected that the negation regions would be separated in this way.

The bond orders can now be obtained, and in this case the two ends are identical for the three species so we expect to find $B_{12} = B_{23}$. Using the populations of the orbitals shown in Figure 11.6 and the coefficients of the wave functions, we find π bond orders as

$$\begin{aligned}
\text{cation } (B_{12} = B_{23})&: 2(0.500)(0.707) = 0.707 \\
\text{radical } (B_{12} = B_{23})&: 2(0.500)(0.707) + 1(0)(0.707) = 0.707 \\
\text{anion } (B_{12} = B_{23})&: 2(0.500)(0.707) + 2(0)(0.707) = 0.707.
\end{aligned}$$

Since the difference in electron populations for the three species involves an orbital of energy α, it is an orbital having the same energy as the atomic orbital and it is therefore nonbonding. Thus, the number of electrons in this orbital (0, 1, or 2 for the cation, radical, or anion, respectively) does not affect the bond orders of the species. The bond orders are determined by the population of the bonding orbital, which has $E = \alpha + \sqrt{2}\beta$.

If the structures C=C–C and C$\cdots$C$\cdots$C are considered, we find that the first structure has one π bond between adjacent carbon atoms, as did ethylene. If we take that bond to be localized, the orbital energy for two electrons would be $2(\alpha + \beta)$ as it was for the ethylene case. For the second structure, the energy that we found for the orbital having the lowest energy is that for the bonding orbital in the allyl system. Therefore, for two electrons populating that orbital, the energy would be $2[\alpha + (2)^{1/2}\beta]$. The difference between these energies is known as the delocalization or resonance energy and amounts to -0.828β. Therefore, the structure showing the delocalized π bond represents a lower energy.

If three carbon atoms are placed in a ring structure, carbon atom 1 is bonded to carbon atom 3 and the secular determinant must be modified to take into account that bond:

C_3

C_1—C_2

Therefore, the elements H_{13} and H_{31} rather than being zero are set equal to β, and the secular determinant is written as

$$\begin{vmatrix} x & 1 & 1 \\ 1 & x & 1 \\ 1 & 1 & x \end{vmatrix} = 0. \tag{11.82}$$

Therefore, we find that the characteristic equation is

$$x^3 - 3x + 2 = 0. \tag{11.83}$$

The roots of this equation are $x = -2$, 1, and 1, which correspond to the energies $E = \alpha + 2\beta$ and $E = \alpha - \beta$ (twice). The resulting the molecular orbital diagram can be represented as shown in Figure 11.7. The coefficients of the wave functions, electron densities, and bond orders can be calculated by the procedures employed for the allyl system. However, we will leave that as an exercise and progress directly to the determination of the resonance energy. For the cation, two electrons populate the lowest state and they have a total energy of $E = 2(\alpha + 2\beta)$, compared with two electrons in an ethylene (localized) bond that have an energy of $E = 2(\alpha + \beta)$. Therefore, the resonance energy is -2β. For the anion, the two degenerate orbitals having an energy of $\alpha - \beta$ are singly occupied and the total energy for four electrons is $E = 2(a + 2\beta) + 2(a - \beta)$. For one

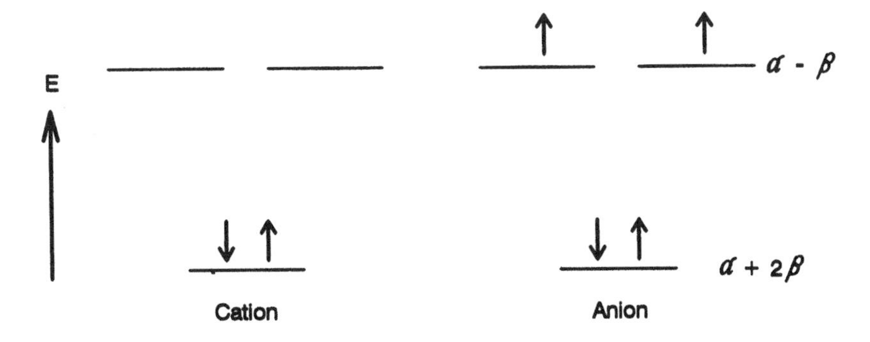

Figure 11.7 ▶ Energy level diagrams for cyclopropenyl species.

localized bond as in ethylene and two single electrons on carbon atoms ($E = \alpha$), the total energy would be $E = 2(\alpha + \beta) + 2\alpha$. Therefore, the resonance energy for the anion would be 0. We correctly predict that the cyclopropenyl cation (resonance energy of 2β) would be more stable than the anion.

Having solved the problems of three carbon atoms in a chain or ring structure, we could use the same methods to examine a totally different chemical system. Suppose we have the H_3^+ species and we wish to determine whether

$$[H_1{-}H_2{-}H_3]^+ \quad \text{(A)} \qquad \textit{or} \qquad \left| \begin{array}{c} H_3 \\ H_1 \text{—} H_2 \end{array} \right|^+ \quad \text{(B)}$$

is more stable. We can still use α as the binding energy of an electron in an atom, but it represents the binding energy of an electron in a hydrogen atom. Therefore, the secular determinants for the two cases will look exactly as they did for the allyl and cyclopropene systems. For structure A, the roots of the secular determinant lead to $E = \alpha + (2)^{1/2}\beta$, $E = \alpha$, and $E = \alpha - (2)^{1/2}\beta$. For structure B, the roots of the secular determinant lead to the energies $E = \alpha + 2\beta$, $E = \alpha - \beta$, and $E = \alpha - \beta$. Therefore, with two electrons to place in the sets of molecular orbitals, we place them in the lowest energy level. For the ring structure that corresponds to $E = 2(\alpha + 2\beta)$ but for the linear structure the energy would be $E = 2[\alpha + (2)^{1/2}\beta]$. The energy is more favorable for the ring structure, and we predict that it would be more stable by -1.2β.

11.5 Dealing with Heteroatoms

In the treatment of organic molecules by the Hückel method, the two energy parameters are H_{11} or α and H_{12} or β. These energies refer to carbon atoms and specifically to the p orbitals of the carbon atom. If atoms other than carbon are present, the Hückel method can still be used, but the energy parameters must be adjusted to reflect the fact that an atom other than carbon has a different binding energy for its electron. Pauling and Wheland developed a procedure to adjust for the heteroatom by relating the α and β parameters for that atom to those for carbon. For example, if a nitrogen atom is present and donates one electron to the π system, the atom is represented in the secular determinant by $\alpha_{N1} = \alpha + 0.5\beta$, where α and β are the values used for a carbon atom, and the subscript "1" indicates the number of electrons used in π bonding. When an atom like nitrogen or oxygen donates more than one electron to the π system, the parameters are adjusted to reflect this difference. A series of values have been adopted for various atoms other than carbon and their adjusted values are shown in Table 11.1.

Earlier, we discussed the allyl system as an example of a molecule containing three atoms. To illustrate the type of calculation possible for systems containing atoms other than carbon, we will consider the N–C–N molecule, in which it is assumed that each atom contributes one electron to the π system. In this case, we will use $\alpha_{N1} = \alpha + 0.5\beta$, and without introducing any other changes, the secular determinant can be written as

$$\begin{vmatrix} \alpha + 0.5\beta - E & \beta & 0 \\ \beta & \alpha - E & \beta \\ 0 & B & \alpha + 0.5\beta - E \end{vmatrix} = 0. \tag{11.84}$$

TABLE 11.1 ▶ Values of Coulomb and Exchange Integrals for Heteroatoms

$\alpha_{N1} = \alpha + 0.5\beta$
$\alpha_{N2} = \alpha + 1.5\beta$
$\alpha_{O1} = \alpha + 1.5\beta$
$\alpha_{O2} = \alpha + 2.5\beta$

Note. The number following the symbol for the atom indicates the number of electrons used in π bonding.

As before, we divide each element by β and let $x = (\alpha - E)/\beta$. The secular determinant then becomes

$$\begin{vmatrix} x+0.5 & 1 & 0 \\ 1 & x & 1 \\ 0 & 1 & x+0.5 \end{vmatrix} = 0. \tag{11.85}$$

This determinant can be expanded to give

$$x(x+0.5)^2 - (x+0.5) - (x+0.5) = 0 \tag{11.86}$$

or

$$x^3 + x^2 - 1.75x - 1 = 0. \tag{11.87}$$

The roots of this equation are $x = -1.686$, -0.500, and 1.186, which lead to the energy levels shown in Figure 11.8.

The coefficients in the wave functions can be evaluated in the usual way since from the secular determinant we can write the equations

$$a_1(x+0.5) + a_2 = 0 \tag{11.88}$$

$$a_1 + a_2x + a_3 = 0 \tag{11.89}$$

$$a_2 + a_3(x+0.5) = 0. \tag{11.90}$$

From Eqs. (11.88) and (11.90) we can see that $a_1 = a_3$. Starting with the root $x = -1.686$, we find that $x + 0.5 = -1.186$ and after substituting $a_1 = a_3$ the equations become

$$-1.186a_1 + a_2 = 0 \tag{11.91}$$

$$a_1 - 1.686a_2 + a_1 = 0 \tag{11.92}$$

$$a_2 - 1.186a_1 = 0. \tag{11.93}$$

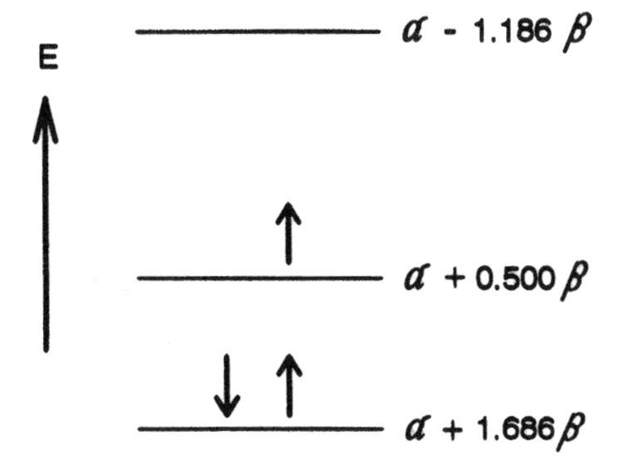

Figure 11.8 ▶ Energy level diagram for the NCN molecule. Each nitrogen atom donates one electron.

We now see that $a_2 = 1.186a_1 = 1.186a_3$ and from the normalization condition we can see that

$$a_1^2 + a_2^2 + a_3^2 = 1 = a_1^2 + (1.186a_1)^2 + a_1^2. \tag{11.94}$$

Therefore, we find $a_1 = (1/3.41)^{1/2} = 0.542 = a_3$, which results in $a_2 = 1.186a_1 = 0.643$. Accordingly, the wave function that corresponds to the root $x = -1.686$ is

$$\psi_1 = 0.542\phi_1 + 0.643\phi_2 + 0.542\phi_3. \tag{11.95}$$

Using the $x = -0.500$ root in an analogous procedure leads to the wave function

$$\psi_2 = 0.707\phi_1 - 0.707\phi_3. \tag{11.96}$$

Since these two orbitals for which we have obtained wave functions can hold the three electrons in the π system, we will not need to determine the coefficients for the third molecular orbital.

A convenient way to determine whether the calculated energy expressions are correct is to substitute the wave function in the equation

$$E = \int \psi_1^* \hat{H} \psi_1 \, d\tau. \tag{11.97}$$

If the calculations are correct, the integral must evaluate to the energy found, $E = a + 1.686\beta$. Substituting the wave function ψ_1 into this expression gives

$$\begin{aligned} E_1 &= a_1^2(\alpha + 0.5\beta) + a_2^2\alpha + a_3^2(\alpha + 0.5\beta) \\ &\quad + 2a_1a_2\beta + 2a_2a_3\beta \\ &= 0.542^2(\alpha + 0.5\beta) + (0.643)^2\,\alpha + (0.542)^2\,(\alpha + 0.5\beta) \\ &\quad + 2(0.643)(0.542)\beta + 2(0.542)(0.643)\beta, \end{aligned} \tag{11.98}$$

which simplifies to $E_1 = 1.00\alpha + 1.686\beta$, which is the energy calculated earlier. Therefore, the calculated energy is correct. We could also verify the value of E_2 by a similar procedure.

The electron densities for N_1–C_2–N_3 (where subscripts indicate atomic positions in the chain) are

$$\text{ED at } N_1 = 2(0.542)^2 + 1(0.707)^2 = 1.09$$

$$\text{ED at } C_2 = 2(0.643)^2 = 0.827$$

$$\text{ED at } N_3 = 2(0.542)^2 + 1(-0.707)^2 = 1.09.$$

Note that to within the round-off errors encountered, the sum of the electron densities is 3.0. As expected based on the relative electronegativities of nitrogen and carbon, we find that the electron density is higher on the nitrogen atoms than it is on the carbon atom.

The bond orders $B_{12} = B_{23}$ can be calculated as illustrated earlier:

$$B_{12} = B_{23} = 2(0.542)(0.643) + 1(0.707)(0) = 0.697. \tag{11.99}$$

If we assume that the nitrogen atoms each donate two electrons to the π system, the value $\alpha_{N2} = \alpha + 1.5\beta$ is used, and after making the usual substitutions the secular determinant can be written as

$$\begin{vmatrix} x+1.5 & 1 & 0 \\ 1 & x & 1 \\ 0 & 1 & x+1.5 \end{vmatrix} = 0. \tag{11.100}$$

This results in the characteristic equation

$$x(x+1.5)^2 - (x+1.5) - (x+1.5) = 0, \tag{11.101}$$

which has the roots $x = -2.35$, -1.50, and $+0.85$. These roots are then set equal to $(\alpha - E)/\beta$, and the calculated energies lead to the molecular orbital diagram shown in Figure 11.9.

Using the procedures developed earlier, we find the coefficients to be $a_1 = 0.606$, $a_2 = 0.515$, and $a_3 = 0.606$ when the root $x = -2.35$ (corresponding to the lowest energy) is used. Therefore, the lowest lying molecular orbital has the wave function

$$\psi_1 = 0.606\phi_1 + 0.515\phi_2 + 0.606\phi_3.$$

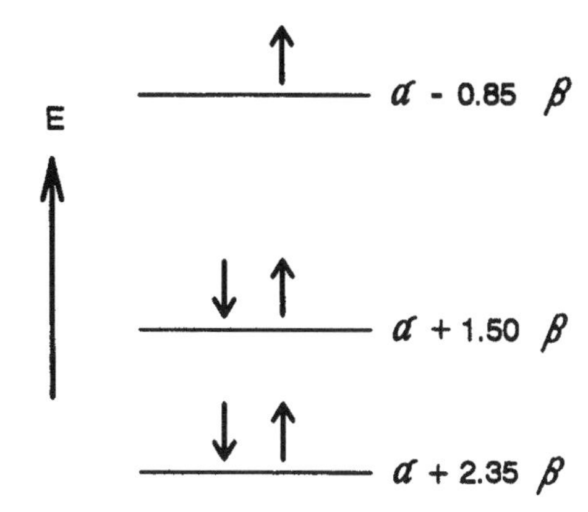

Figure 11.9 ▶ Energy level diagram for the NCN molecule. Each nitrogen atom donates two electrons.

From the root $x = -1.50$, we find the coefficients $a_1 = 0.707$, $a_2 = 0$, and $a_3 = -0.707$, which lead to the wave function

$$\psi_2 = 0.707\phi_1 - 0.707\phi_3.$$

The coefficients $a_1 = 0.365$, $a_2 = -0.857$, and $a_3 = 0.365$ are found from the root $x = 0.85$. That root gives rise to the wave function

$$\psi_3 = 0.365\phi_1 - 0.857\phi_2 + 0.365\phi_3,$$

which corresponds to the orbital of highest energy, which is populated by only one electron. When the coefficients of the three wave functions are used to calculate the electron densities on the atoms, the results are

$$\text{ED at } N_1 = 2(0.606)^2 + 2(0.707)^2 + 1(0.365)^2 = 1.87$$
$$\text{ED at } C_2 = 2(0.515)^2 + 2(0)^2 + 1(0.857)^2 = 1.26$$
$$\text{ED at } N_3 = 2(0.606)^2 + 2(0.707)^2 + 1(0.365)^2 = 1.87.$$

As expected, the total electron density is equivalent to five electrons, and the electron density is higher on the nitrogen atoms than it is on the carbon atom.

It should be mentioned that the cyanamide ion, CN_2^{2-}, has the structure N=C=N and constitutes a logical extension of the preceding problem. It is thus apparent that possibilities exist for using the Hückel molecular orbital calculations for inorganic species as well. The main use of this method has, of course, been in the area of organic chemistry.

11.6 Orbital Symmetry and Reactions

In Chapter 9, the applications of orbital symmetry to the formation of transition states between reacting diatomic molecules were described. Having now shown the applications of Hückel molecular orbital theory to the structures of organic molecules, we are now in a position to apply some of the principles to reactions of organic compounds. One type of reaction that can be described in terms of orbital symmetry is the ring-closing reaction of *cis*-1,3-butadiene to produce cyclobutene. This type of reaction is known as an *electrocyclic reaction*, and it could conceivably take place by two different pathways, which are shown in Figure 11.10. In the first mechanism, known as *conrotatory*, the two CH_2 groups rotate in the same direction. In the second mechanism, known as *disrotatory*, the terminal CH_2 groups rotate in opposite directions. However, the stereochemistry of the product will be different for the two mechanisms.

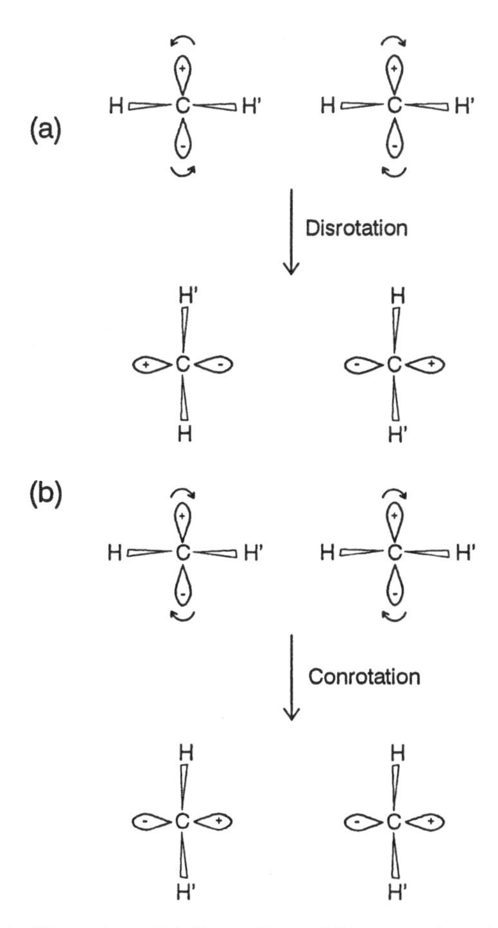

Figure 11.10 ▶ An illustration of (a) disrotation and (b) conrotation of terminal methylene groups.

If the reaction follows a disrotatory pathway, the rotation of terminal CH_2 groups is as shown in Figure 11.10(a) while that for the conrotatory pathway is shown in Figure 11.10(b). In order to show the stereochemistry of the product for each of the two pathways, one hydrogen atom on each terminal CH_2 group is labeled as H'. It is clear that the hydrogen atoms labeled as H' would be on the same side of the ring in a conrotatory process and on opposite sides of the ring in a disrotatory pathway. Let us now show how orbital symmetry can be used to predict which transition state will be formed.

A pictorial representation of the HOMO for butadiene is shown in Figure 11.11. From this figure, it can be seen that rotation of the CH_2 groups in a conrotatory pathway leads to overlap of orbitals having the

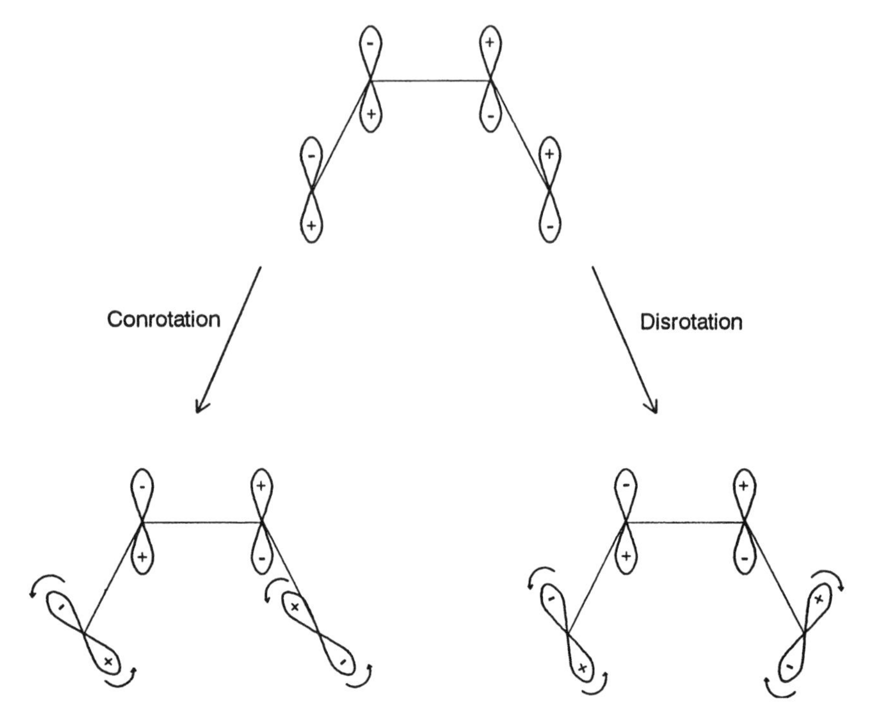

Figure 11.11 ▶ Symmetry of the HOMO of 1,3-butadiene showing changes in orbital orientation during conrotation and disrotation.

same symmetry (bond formation) and the product formed has the H′ atoms on opposite sides of the ring. In the disrotatory process, the overlap of orbitals of opposite symmetry results, which gives an overlap of 0. In accord with these observations, the product of the electrocyclic reaction that results when 1,3-butadiene is heated consists of 100% of that in which the H′ atoms are on opposite sides of the ring. Of course the reaction can also be studied when one of the hydrogen atoms on the terminal CH_2 groups is replaced by a different atom. In that case, the substituted atoms are found on opposite sides of the ring in the cyclobutene produced.

When 1,3-butadiene is excited photochemically, an electron is moved from the HOMO to the LUMO, which has different symmetry (see Figure 11.12). Therefore, it is the disrotatory motion that brings orbital lobes having the same symmetry (sign) in contact as the bond forms during the cyclization reaction of the excited state of 1,3-butadiene. As a result, the product of this reaction of excited 1,3-butadiene is found to have the H′ atoms on the same side of the ring.

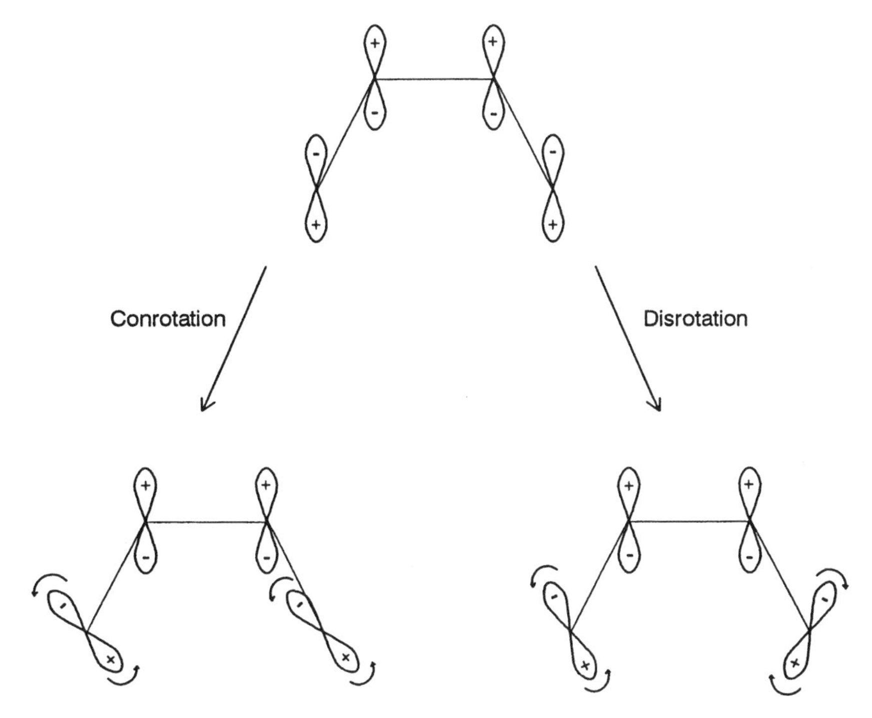

Figure 11.12 ▶ Symmetry of the LUMO of 1,3-butadiene showing changes in orbital orientation during conrotation and disrotation.

The electrocyclic ring closure of cis-1,3,5-hexatriene leads to the formation of 1,3-cyclohexadiene. Although the hexatriene molecule is planar, the cyclic product has two CH_2 groups in which the four hydrogen atoms are found with two of the atoms above the ring and two of the atoms below the ring. Therefore, in the transition state the terminal CH_2 groups undergo a rotation that could be either conrotatory or disrotatory as shown in Figure 11.13. The rotations require breaking of a π bond formed from p orbitals on two carbon atoms to form a σ bond. In order to give positive overlap to form the σ bond, the orbitals must match in symmetry. This match is provided by the disrotatory pathway. Therefore, in disrotatory motion, the rotation of the terminal CH_2 groups leads to the formation of a bond between the two carbon atoms resulting in ring closure. In the conrotatory motion of the CH_2 groups, the resulting orbital overlap would be zero.

A guiding principle that can be used to predict how electrocyclic reactions take place was provided by R. B. Woodward and R. Hoffmann. This rule is based on the number of electrons in the n bonding system of

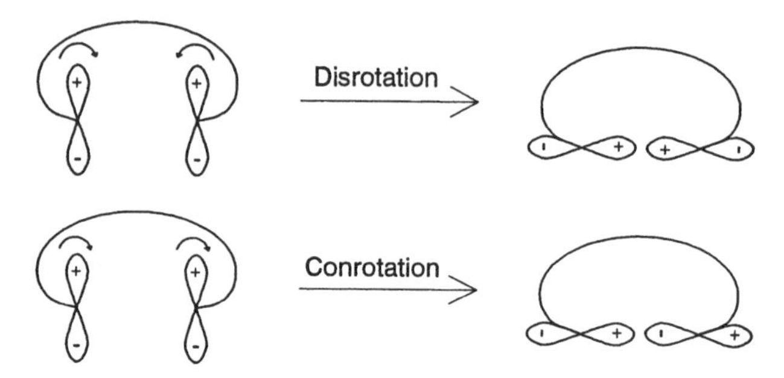

Figure 11.13 ▶ Disrotation and conrotation of terminal methylene groups in 1,3,5-hexatriene. The rest of the molecule is indicated by the line.

the molecule. The number of electrons in the π bonding system can be expressed as $4n$ or $4n + 2$, where $n = 0, 1, 2, \ldots$. The rule predicts the mechanism of electrocyclization as

$4n = 4, 8, 12, \ldots$, thermal mechanism is conrotatory.
$4n + 2 = 2, 6, 10, \ldots$, thermal mechanism is disrotatory.

It can be seen that the electrocyclization reactions of 1,3-butadiene (which has four electrons in p orbitals) and 1,3,5-hexatriene (which has six electrons in p orbitals) discussed equation give products in accord with these rules. However, it must be mentioned that the predictions given above are for the reactions that are thermally induced. If the reactions are carried out photochemically where excited status are produced, the predictions are reversed.

The discussion presented here has provided only an introduction to how symmetry of molecular orbitals can be used to predict reaction mechanisms. It should be apparent that the requirement that the transition state involves interaction of orbitals of like symmetry is a powerful tool for predicting reaction pathways. Chemists doing synthetic work as well as theoretical chemists need to be familiar with these aspects of molecular orbital theory. However, a knowledge of only Hückel molecular orbital theory is required in order to deal with many significant problems in molecular structure and reactivity. In view of its gross approximations and very simplistic approach, it is surprising how many qualitative aspects of molecular structure and reactivity can be dealt with using the Hückel approach. For a more complete discussion of this topic, consult the book by J. P. Lowe cited in the references.

References for Further Reading

- Adamson, A. W. (1986). *A Textbook of Physical Chemistry*, 3rd ed., Chap. 18. Academic Press College Division, Orlando. A good introduction to the quantum mechanics of bonding.
- Drago, R. S. (1992). *Physical Methods for Chemists*, Chap. 3. Saunders College Publishing, Philadelphia. Results from extended Hückel and other molecular orbital calculations.
- Liberles, A. (1966). *Introduction to Molecular-Orbital Theory*. Holt, Rinehart, & Winston, New York. An introductory level book with several problems solved in detail.
- Lowe, J. P. (1993). *Quantum Chemistry*, 2nd ed. Academic Press, San Diego. An excellent treatment of quantum chemistry at a rather high level. Thorough coverage of molecular orbital theory applied to reactions.
- O'Neil, P. V. (1993). *Advanced Engineering Mathematics*, 3rd ed. PWS Publishing, Boston, A thorough coverage of almost all advanced mathematical topics including differential equations and determinants.
- Riggs, N. V. (1969). *Quantum Chemistry.* Macmillan, New York. A solid book on quantum chemistry with an excellent treatment of bonding.
- Roberts, J. D. (1962). *Notes on Molecular Orbital Calculations.* Benjamin, New York. The classic book that had much to do with popularizing molecular orbital calculations.
- Wolfsberg, M., and Helmholtz, L. (1952). *J. Chem. Phys.* **20**, 837. The original publication dealing with the Wolfsberg–Helmholtz approximation.

Problems

1. Find the roots of these equations using graphical or numerical methods:

(a) $x^3 - 2x^2 - 2x + 3 = 0$,

(b) $x^4 - 5x^2 + 2x - 5 = 0$,

(c) $x^3 + 2x^2 - 6x - 8 = 0$,

(d) $x^6 - 6x^4 + 9x^2 - 4 = 0$.

2. Find the solution to $e^x + \cos x = 0$, where x is in radians.

3. Solve $\exp(-x^2) - \sin x = 0$ if x is in radians.

4. Find two roots for the equation $e^{2t} - 2\cos 2t = 0$.

5. In the text, the Hückel molecular orbital calculations were carried out for H_3^+. Perform a similar analysis for H_3^-.

6. Two possible structures of I_3^+ are

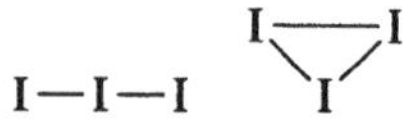

I—I—I   (triangle of three I atoms)

Use a Hückel molecular orbital orbital calculation to determine which structure is more likely.

7. The tri-iodide ion, I_3^-, forms when I_2 reacts with I^- in aqueous solutions of KI. Perform the same calculations as in Problem 6 for the I_3^- to determine the preferred structure.

8. Perform Hückel molecular orbital calculations for the molecules N–C–C and C–N–C. In each case, assume that the nitrogen atom is a two-electron donor to the π system. Determine the energy levels, the coefficients for the wave functions, and the charge densities on the atoms.

9. Perform Hückel molecular orbital calculations for bicyclobutadiene,

C C C C

10. Carry out a Hückel molecular orbital calculation for 1,3-butadiene, $H_2C{=}CH{-}CH{=}CH_2$. Determine the energy levels, the coefficients of the wave functions, the bond orders, and the electron density at each carbon atom.

11. Using the Hückel molecular orbital approach, determine the resonance or delocalization energy for cyclobutadiene,

C=C
| |
C=C

Complete the calculations by determining the coefficients of the wave functions, the electron density on each atom, and the bond orders.

12. Repeat the calculations of Problem 11 for the ring structure

C=N
| |
N=N

in which each nitrogen atom is assumed to contribute one electron to the π system.

13. Repeat the calculations of Problem 11 for a ring structure like

```
N=N
|  |
C=C
```

assuming that each nitrogen atom contributes one electron to the π system.

14. Consider the molecule

```
C—C—C
‖  ‖  ‖
C—C—C
```

Show how the symmetry of this molecule could be used to simplify the Hückel calculations.

15. Assume that a linear structure having a -1 charge is composed of one atom of I, one of Cl, and one of Br. Based on the chemical nature of the atoms, what should be the arrangement of atoms in the structure? Consider only the p orbitals and assume that the same α value can be used for each of these atoms (a very crude approximation). Perform Hückel calculations to determine the electron density at each atom and the bond orders.

16. Because Cl, Br, and I atoms have different electronegativities and electron binding energies (ionization potentials), the same value for α should not be used for each atom. Use the values α for I, $\alpha + 0.2\beta$ for Br, and $\alpha + 0.4\beta$ for Cl to compensate for the difference in properties of the atoms. Perform Hückel calculations for the three possible linear arrangements of atoms and calculate the electron densities on the atoms for each structure. Explain how the results of the calculations support or contradict the structure that you would expect based on the chemical nature of the atoms.

▶ Chapter 12

More Complete Molecular Orbital Methods

In any area of science in which the results are important, there will be a great deal of original thinking and research. Such is the case for molecular orbital calculations. As a result, a large number of approaches to performing molecular orbital calculations exist. They are known by various acronyms that may be unintelligible to all except those who have devoted serious study to the field. It is not possible in one small book (or one large one for that matter) to present the details of this enormous body of knowledge. Consequently, in this chapter the attempt will be to present an overview of the language and a qualitative understanding of some of the most important techniques in this important field.

While the simple Hückel method (HMO) described in the previous chapter is useful for some purposes, that type of calculation is quite limited and is most applicable to organic molecules. More sophisticated types of calculations require enormously more complex computational techniques. As a result, the developments in the field of molecular orbital calculations have paralleled the developments in computers. Computer software is now routinely available from several sources that enables persons who do not necessarily understand all of the theory or computational techniques to load the software and be guided through the process of performing high-level molecular orbital calculations that would have represented the frontier of the field not many years ago. Because of the computer system requirements, specific instructions depend on the type of computer equipment on which the calculations are to be performed. In view of these aspects of molecular orbital methods, the discussion in this chapter will be limited to presenting basic principles and nomenclature.

12.1 The Basis Set

All of the molecular orbital calculations make use of some type of atomic wave functions that generally describe a single electron. The *basis set* is the set of one-electron wave functions that will be combined to give the molecular wave functions. The minimal basis set is the set that incorporates only the orbitals actually populated by electrons. Slater-type orbitals (STO) having the form

$$\psi(r) = r^{n-1}\mathrm{e}^{-(Z-s)r/n}, \tag{12.1}$$

where s is a screening constant, n is a number that varies with the type of orbital, and Z is the nuclear charge, are widely used. The value of n is determined according to procedures described in Section 5.2. Using one STO wave function for each nucleus and constructing molecular wave functions by taking linear combinations of the atomic orbitals, the molecular calculation is referred to as the minimal basis set calculation. The quantity $(Z - s)$ is replaced by ζ (zeta) to give functions written in the form

$$\psi(r) = r^{n-1}\mathrm{e}^{-\zeta r/n}. \tag{12.2}$$

The wave functions of this type are referred to as single-ζ STO functions. An additional modification involves representing each atomic wave function by two STO wave functions. In this case, the wave functions are referred to as double-ζ functions. Within the framework of double-ζ wave functions, there are several variations in parameterizations that have been developed.

Another type of function used to represent atomic wave functions is known as a Gaussian. Gaussian functions, which are in the form $\psi(r) = \exp(-\alpha r^2)$ where α is an adjustable (best-fit) parameter, are combined as linear combinations in order to approximate the STO functions. The motivation for this is that computations of the integrals involved in the quantum mechanical calculations are greatly facilitated. The linear combination of Gaussian functions is referred to as a contracted Gaussian function or Gaussian-type orbitals (GTO). All of these manipulations of STO and GTO are to provide approximations to the radial portions of the atomic functions, and the complete wave functions are obtained by making use of the spherical harmonics, $Y_{l,m}(\theta, \phi)$, to provide the angular dependence.

When each STO is represented by a linear combination of three Gaussian functions, the result is known as STO-3G. Other combinations of wave functions lead to the 6-31G designation in which each STO is represented as a linear combination of six Gaussian functions. Further, each STO representing valence shell orbitals is a double-ζ function with the inner part represented by a linear combination of three Gaussian functions and the outer by one such function. Although the description of the types of functions present is by no means complete, it does show that many creative mathematical approaches have been utilized.

12.2 The Extended Hückel Method

In the previous chapter, we have illustrated the applications of the Hückel method to a variety of problems. In 1963, Roald Hoffmann (Nobel prize, 1981) devised a molecular orbital method that has come to be known as the extended Hückel molecular orbital method (EHMO).[1] This method has several differences from the basic Hückel method. For organic molecules, the basis set of carbon $2s$ and $2p$ and hydrogen $1s$ orbitals is used in the calculations. Although, the overlap was neglected in the Hückel method, it is explicitly included in the EHMO procedure. All overlap integrals, S_{ij}, must be calculated. Since the atomic wave functions are normalized, the S_{ii} integrals are equal to 1.

The values of the overlap integrals depend on bond distances and angles. Therefore, before the overlap integrals can be evaluated, the relative positions of the atoms must be known. In other words, the results of the calculation will depend on the molecular geometry. If one wishes to determine the effect of changing bond angles or distances, these parameters can be changed and the calculation repeated. The choice of a coordinate system must be made so that the positions of the atoms can be calculated. For example, if a calculation were to be performed for the trigonal planar BH_3 molecule, the coordinates might be set up so that the boron atom is at the origin, one hydrogen atom lies on the x axis, and the other two lie in the xy plane between the x and y axes. After the coordinates of the atoms are determined, the overlap integrals can be evaluated (see Section 9.5).

In a calculation for an organic molecule, the basis set consists of $1s$ wave functions for the hydrogen atoms and the $2s$ and $2p$ wave functions for the carbon atoms. Thus, for C_nH_m, the basis set consists of m hydrogen $1s$ wave functions and n $2s$ and $3n$ $2p$ carbon orbitals. Slater-type orbitals are most commonly chosen with the exponents determined as outlined in Section 5.2. The procedures lead to a value of 1.625 for carbon orbitals

[1] See the references at the end of this chapter.

and 1.0 for hydrogen orbitals although the values may vary somewhat. In the original work, Hoffmann used a value of 1.0 but other workers have suggested a value of 1.2 may be more appropriate (for example, see Lowe, Chapter 10). Since the atomic wave functions have been deduced, the values for the overlap integrals can now be computed. This produces the values that make up the overlap matrix.

The Coulomb integrals, H_{ii}, are approximated as the valence state ionization potentials for removal of an electron from the orbital being considered. For the hydrogen atom, the ionization potential is 13.6 eV so the binding energy for an electron in a hydrogen atom is taken as -13.6 eV. For organic molecules, which were the subject of Hoffmann's original work, the choice of ionization potentials is not so obvious. In many organic molecules, the carbon is hybridized sp^2 or sp^3 so the loss of an electron from a carbon $2s$ or $2p$ orbital does not correspond exactly to the binding energy of an electron in a carbon atom in a molecule. Therefore, there is some choice to be made as to what value is to be used for the ionization potential.

We saw in Chapter 11 that the H_{ij} integrals were set equal to 0 when $|i - j| > 1$. In other words, only interactions between adjacent atoms were included. Unlike the simple Hückel method, the off-diagonal elements of the Hamiltonian matrix, the H_{ij}, are not omitted in the EHMO regardless of the positions in the molecule. The Wolfsberg–Helmholtz approximation,

$$H_{ij} = KS_{ij}\frac{H_{ii} + H_{jj}}{2}, \tag{12.3}$$

is ordinarily used to compute the values of the integrals. Having determined the values for the H_{ii} and H_{ij} integrals, the matrix that gives the values for the energy integrals is constructed. As in the case of the overlap matrix, the dimension is equal to the total number of atomic orbitals included in the basis set. Once the overlap and Hamiltonian matrices (S and H) are obtained, the equation to be solved in matrix form is $\mathsf{HC} = \mathsf{SCE}$, where E is the energy eigenvalue. These are computations performed by a computer.

While the EHMO method is not equivalent to SCF calculations, it is still a useful method for certain types of problems. For example, calculated energy barriers for rotation, such as the difference between the staggered and eclipsed conformations of ethane, agree reasonably well with experimental values. The EHMO method has also proved useful for calculations on extended arrays as in the case of solids and semiconductors. As was shown in Chapter 10, the HMO method can give insight into the behavior of organic molecules. The EHMO method is even better at the expense of greater computational effort.

12.3 The HF-SCF (Hartree–Fock Self-Consistent Field) Approach

In Chapter 5, the helium atom was the subject of two approximation methods of great importance when treating problems that cannot be solved exactly. The difficulty was the $1/r_{ij}$ term in the Hamiltonian, which arises because of repulsion between electrons, that prevented the separation of variables. When the variation method was applied, it was found that repulsion between the two electrons in the helium atom caused them to "see" a nuclear charge smaller than +2 with the effective nuclear charge being 27/16. Using that value, the calculated ground state energy was rather close to the experimental value.

In the second approach, the presence of a second electron was considered as a perturbation on the behavior of the other. In that case, the calculated perturbation energy was $(\frac{5}{4})ZE_H$, which leads to a ground state energy that is also rather close to the experimental value. Although the variation and perturbation methods are manageable for an atom that has two electrons, the approach becomes exceedingly complex when the atom being studied has a sizable number of electrons because of the large number of terms involving $1/r_{ij}$ when all of the interactions are included. A method for dealing with such complex calculations was developed by D. R. Hartree, who expanded the procedures of V. Fock. The result is a type of calculation known as the *Hartree–Fock* or *self-consistent field* (SCF) calculation.

In principle, the SCF approach to calculations on atoms is rather straightforward. In the case of the helium atom, it was assumed that each electron moved around the nucleus in much the same way as the electron does in a hydrogen atom. In the SCF approach, each electron is assumed to move in a spherical central electrostatic field generated by all of the other electrons and the nucleus. Filled shells do, in fact, generate a spherically symmetric field, and the nature of the calculation averages somewhat the effects of fields that deviate from spherical character. The calculation is carried out for the electron in question as it is acted on by the field generated by the nucleus and other electrons. The hydrogen-like wave functions are utilized in calculating the spatial distribution of the field. Carrying out the calculation for each electron, the behavior of the electrons generating the field leads to a new, calculated charge distribution. The calculated field is then used as the basis for a new calculation of the behavior of each electron to produce a second charge distribution or field. This is then used to calculate the behavior of each electron in the field, which leads to a new charge distribution.

The process can be carried out until the calculated charge distribution in the field after the nth step is identical with that from the $(n-1)$th step. At that point, additional calculations do not lead to an improved "field" and the field is said to be "self-consistent."

For a calculation based on the helium atom, the procedure considers electron 1 to be moving in a field determined by the nucleus surrounded by the negative charge cloud produced by electron 2. The result is that electron 1 will not move in the same pattern that it would if electron 2 were absent so the calculated wave function for electron 1 will be somewhat different in coefficient and exponent than it would be for a strictly hydrogen-like wave function. With the new parameters determined, the effect on electron 2 can be treated in a similar fashion with the improved wave function. The process is repeated until there is no improvement in the wave functions and the field is considered to be self-consistent.

Unlike the Hückel method described in Chapter 11, SCF calculations involve the evaluation of all integrals and are therefore known as *ab initio* calculations. In SCF calculations, the results obtained will be somewhat dependent on the basis set used. Any basis set chosen is a set of approximate wave functions so the calculated energy will be higher than the true energy (see Section 4.4, p. 70). After a SCF calculation is performed, the results can be used to expand the basis set, which will enable an improved calculation to be made. Further changes can be made in the basis set and the calculation can be repeated. At some point, the calculated energy will approach a minimum value such that changes in the basis set do not result in a lower value for the calculated energy. The lowest energy calculated is known as the Hartree–Fock energy for that system.

According to the variation theorem, the Hartree–Fock energy will still be higher than the true energy. In the SCF method, it assumed that each electron moves in a spherically symmetric field generated by the presence of other electrons in the proximity. In other words, an electron moves in a symbiotic way with other electrons. Therefore, the motion of the electrons are said to be *correlated*. Because this type of interaction is not accounted for in a Hartree–Fock type of calculation, the actual energy will be lower than that calculated because the electrons moving in correlated ways lowers the energy of the system. A procedure known as configuration interaction has been developed to include the effects of electron correlation. The essential idea behind the method is that a linear combination of wave functions is used in which *each determinant* represents a wave function that incorporates electron permutations. This means that for a simple molecule like H_2, electron 1 can be found near nucleus 1 with a spin α, electron 2 can be found near nucleus 2 with a spin β, electron 2 can be found near nucleus 1 with a spin β, etc. The wave function for the ground state ($1\sigma_g^2$) of the H_2 molecule can be written in determinant form as

$$\psi = \frac{1}{2^{1/2}} \begin{vmatrix} 1s(1)\alpha(1) & 1s(1)\beta(1) \\ 1s(2)\alpha(2) & 1s(2)\beta(2) \end{vmatrix}. \tag{12.4}$$

This type of determinental wave function is known as a *Slater determinant.*

The inclusion of other determinantal wave functions as linear combinations leads to an improved basis set. This procedure specifically allows the contributions from excited states to be included. Just as including resonance structures that make even minor contributions to the true structure improves our representation of the structure of a molecule, including wave functions that describe small contributions from excited states leads to an improvement in the energies that result from SCF calculations. For a more complete discussion of configuration interaction and other refinements to the Hartree–Fock SCF calculations, consult the references listed at the end of this chapter, especially the book by Lowe.

One of the problems associated with *ab initio* calculations is that the number of integrals that must be evaluated is very large. In order to provide methods that require calculations on a smaller scale, approaches that ignore certain of the integrals have been developed. In Chapter 5, we saw that one approach to the problem of the helium atom was to ignore the $1/r_{12}$ term in the Hamiltonian. In an analogous way, the interactions between two electrons located in different regions of the molecule are small. Stated another way, the overlap of wave functions for the two electrons is essentially zero. Approximate methods that neglect some overlap integrals are based on the zero differential overlap (ZDO) assumption. The basis for this assumption is that even though wave functions are exponential functions (see Section 4.2, p. 62) and they approach a value of 0 only at an infinite distance, the overlap is *approximately* zero for some orbitals on nonadjacent atoms.

Because of the number of ways in which decisions are made to exclude certain overlap integrals from the calculations, there are several types of approximate MO calculations. Complete neglect of differential overlap, CNDO (which exists in several versions), is one of the early types of approximate computational procedures. Intermediate neglect of differential overlap (INDO) is a computational method that includes overlap of wave functions on the same atom. Other methods neglect the overlap only when the wave functions are for electrons on different atoms. Approximate methods of these types are widely used because they require only limited computing resources and they frequently yield results that are useful for interpreting chemical properties and behavior. In the hierarchy of computational quantum chemistry, they lie somewhere between the Hückel methods on the one hand and the *ab initio* methods on the other.

Although a brief description of some of the molecular orbital methods has been presented in this chapter, further coverage is outside the scope of a book devoted to the fundamentals of quantum mechanics. For more complete discussions of advanced quantum mechanical procedures for molecular systems, consult the references listed at the end of this chapter.

References for Further Reading

- Gavin, R. M. (1969). *J. Chem. Educ.* **46**, 413. An informative tutorial on how to proceed with a molecular orbital calculation to determine the stereochemistry of inorganic molecules.
- Hoffmann, R. (1963). *J. Chem. Phys.* **39**, 1397. The original description of the extended Hückel method.
- Leach, A. R. (1996). *Molecular Modelling*. Pearson Education, Essex. Chapter 2 deals with Hückel, SCF, and semi-empirical methods.
- Levine, I. N. (1974). *Quantum Chemistry*, 2nd ed. Allyn and Bacon, Boston. An older book that provides clear and detailed discussions on many topics dealing with calculations for molecules. Chapters 13 and 15 are especially recommended.
- Lowe, J. P. (1993). *Quantum Chemistry*, 2nd ed. Academic Press, San Diego, CA. Chapter 10 is devoted to the extended Hückel method and Chapter 11 describes SCF methods.

Problem

Find a paper in one of the research journals that publishes work dealing with molecular orbital calculations. Some suggested journals are the *Journal of Chemical Physics*, *Journal of Molecular Structure (THEOCHEM)*, *Journal of Computational Chemistry*, etc. Try to find an article in which as much detail as possible is given about the calculations (not always easy to do). Study the paper thoroughly to determine such things as the choice of basis set, why this particular basis set was chosen, how the molecular coordinates were set up, the type of computer output obtained, how the results were interpreted, etc. After studying the paper carefully, write a summary of about two pages giving an overview of the work. Try to include items that were mentioned in this chapter.

Answers to Selected Problems

Chapter 1

1. 49.5 m.

2. 1.21×10^{-8} cm.

5. $n = 3$ to $n = 41.0$ nm; $n = 4$ to $n = 117$ nm.

7. 490 nm.

9. 1.64×10^{-6} erg.

11. 1.71×10^{-8} cm.

13. 328. It will be less because ejected electrons carry away some kinetic energy.

Chapter 2

1. All are. Eigenvalues are (a) $\hbar$; (b) $l\hbar$; (c) $-\hbar$.

3. (a) $10^{1/2}\, e^{-5x}$; (b) $b^{1/2}\, e^{-bx}$.

7. $\left[\frac{(2b)^{1/2}}{a}\right] e^{-2bx}$.

8. Yes for d^2/dx^2.

Chapter 3

1. $y = A \sin a^{1/2}x + B \cos a^{1/2}x$. Boundary conditions give $y = A \sin a^{1/2}x$.

2. 477 pm assuming the electron behaves as a particle in a box.

3. 1.65×10^{-32} erg; 6.58×10^{-32} erg.

10. 879 nm; such solutions are usually colored blue.

13. 6.

Chapter 4

1. 2.19×10^{8} cm/s.

Chapter 5

1. **(a)** $\frac{11}{2}, \frac{9}{2}, \ldots, \frac{1}{2}$; **(b)** $\frac{7}{2}, \frac{5}{2}, \frac{3}{2}, \frac{1}{2}$.

3. **(a)** $^{2}P_0$; **(b)** $^{4}S_{5/2}$; **(c)** $^{1}S_0$; **(d)** $^{3}F_2$; (e) $^{2}D_{3/2}$.

7. $Z_{\text{eff}} = 5.20$; $(Z - s)/n^* = 2.60$; $\psi = r\, e^{2.60r/a}\, Y_{2,m}(\theta, \phi)$.

9. **(a)** $^{2}D_{3/2}$; **(b)** $^{4}F_{3/2}$; **(c)** $^{1}S_0$; **(d)** $^{6}S_{5/2}$; **(e)** $^{3}F_4$.

Chapter 6

5. **(a)** $y = c_1e^{x} + c_2e^{-x}$; **(c)** $y = -\frac{e^x}{2} + \frac{5}{2}e^{-x}$; **(d)** $y = e^{2x} - 2e^{x}$.

6. **(a)** $y = 1 - x + x^2/2! - x^3/3! + \cdots = e^{-x}$;
(b) $y = 2(1 + x^2/2 + x^4/2^2 2! + x^6/2^3 3! + \cdots) = 2\exp(x^2/2)$;
(c) $y = c_1 \cos x + c_2 \sin x$.

Chapter 7

1. 8.2×10^{13} s^{-1}.

3. 113 pm.

6. 4.82×10^{5} dyne/cm; 4.82 m dyne/Å; 4.82×10^{2} N/m.

7. 129 pm.

9. 7.67 cm^{-1} for $^{12}C^{16}O$ and 7.04 cm^{-1} for $^{14}C^{16}O$.

Chapter 8

1. (**a**) 0.259; (**b**) 0.286; (**c**) 0.319.

3. (**a**) 4.78×10^{-51}; (**b**) 2.58×10^{-47}; (**c**) 2.96×10^{-43}.

5. (**a**) 7.18×10^{-61}; (**b**) approximately 0.

8. (**a**) 9.1×10^{-15}; (**b**) 0.021.

Chapter 9

1. 1.44 D.

3. 1.17 D.

5. 138 pm.

7. 162 pm.

9. (**a**) KK $1\sigma_g^2\ 1\sigma_u^2\ 1\pi_u^4\ 2\sigma_g^1$ B.O. = 2.5;
(**b**) KK $1\sigma_g^2\ 1\sigma_u^2\ 1\pi_u^4$ B.O. = 2;
(**c**) KK LL $1\sigma_g^2$ B. O. = 1;
(**d**) KK LL $1\sigma_g^2\ 1\sigma_u^2\ 2\sigma_g^2\ 1\pi_u^4\ 1\pi_g^2$ B.O. = 2.

12. (**a**) 1.5;
(**b**) $1\pi_u$ is a bonding orbital so dissociation energy decreases;
(**c**) Yes, add one electron to give $1\pi_u^4$ configuration.

14. (**a**) ${}^1\Sigma_g^+$;
(**b**) ${}^1\Sigma_g^+$;
(**c**) ${}^2\Pi_g$;
(**d**) ${}^2\Pi_u$.

16. In the first case, the electron was removed from $2\sigma_g$, but in the second it came from $1\pi_u$.

19. A triplet state indicates that degenerate π_u orbitals have one electron in each. Therefore, they must lie lower in energy than the $2\sigma_g$.

Chapter 10

1. (a) $C_{\infty v}$; (b) C_{2v}; (c) D_{4h}; (d) C_{3v}; (e) O_h; (f) C_{2v}; (g) C_{2v}; (h) D_{3h}; (i) C_{2v}; (j) C_s.

5. For the linear structure, the energy levels are $\alpha + 2^{1/2}\beta$, α, and $\alpha - 2^{1/2}\beta$. For the ring structure, the energy levels are $\alpha + 2\beta$, $\alpha - \beta$, and $\alpha - \beta$. The linear structure is more stable for H_3^-.

7. The results are analogous to problem 5.

9. The bond "across" the ring connects C_2 and C_4. The energy levels are $E_1 = \alpha + 2.56\beta$; $E_2 = \alpha$; $E_3 = \alpha - \beta$; $E_4 = \alpha - 1.56\beta$. The wave functions are

$$\psi_1 = 0.435\phi_1 + 0.557\phi_2 + 0.435\phi_3 + 0.557\phi_4$$
$$\psi_2 = 0.707\phi_1 - 0.707\phi_3$$
$$\psi_3 = 0.707\phi_2 - 0.707\phi_4$$
$$\psi_4 = 0.557\phi_1 - 0.435\phi_2 + 0.557\phi_3 - 0.435\phi_4.$$

Charges on atoms are $q_1 = q_3 = -0.379$ and $q_2 = q_4 = 0.379$.

10. The energy levels are $E_1 = \alpha + 1.62\beta$; $E_2 = \alpha + 0.618\beta$; $E_3 = \alpha - 0.618\beta$; $E_4 = \alpha - 1.62\beta$. The wave functions are

$$\psi_1 = 0.372\phi_1 + 0.602\phi_2 + 0.602\phi_3 + 0.372\phi_4$$
$$\psi_2 = 0.602\phi_1 + 0.372\phi_2 - 0.372\phi_3 - 0.602\phi_4$$
$$\psi_3 = 0.602\phi_1 - 0.372\phi_2 - 0.372\phi_3 + 0.602\phi_4$$
$$\psi_4 = 0.372\phi_1 - 0.602\phi_2 + 0.602\phi_3 - 0.372\phi_4.$$

The π bond orders are (subscripts indicate the carbon atoms) $B_{12} = 0.896 = B_{34}$ and $B_{23} = 0.448$. Add 1 for the σ bond to get total bond orders.

Index